NEUROSCIENCE
INTELLIGENCE
UNIT

CYTOKINES
IN THE NERVOUS SYSTEM

Nancy J. Rothwell

University of Manchester
Manchester, United Kingdom

CHAPMAN & HALL
I⫶P An International Thomson Publishing Company

New York • Albany • Bonn • Boston • Cincinnati • Detroit • London • Madrid • Melbourne •
Mexico City • Pacific Grove • Paris • San Francisco • Singapore • Tokyo • Toronto • Washington

R.G. LANDES COMPANY
AUSTIN

Neuroscience Intelligence Unit

CYTOKINES IN THE NERVOUS SYSTEM

R.G. LANDES COMPANY
Austin, Texas, U.S.A.

Please address all inquiries to the Publishers:
R.G. Landes Company, 909 Pine Street, Georgetown, Texas, U.S.A. 78626
Phone: 512/ 863 7762; FAX: 512/ 863 0081

North American distributor:
Chapman & Hall, 115 Fifth Avenue, New York, New York, U.S.A. 10003

CHAPMAN & HALL

U.S. and Canada ISBN:-0-412-13581-7

Library of Congress Cataloging-in-Publication Data
CIP applied for, but not received as of publication date.

PUBLISHER'S NOTE

R.G. Landes Company publishes six book series: *Medical Intelligence Unit, Molecular Biology Intelligence Unit, Neuroscience Intelligence Unit, Tissue Engineering Intelligence Unit, Biotechnology Intelligence Unit* and *Environmental Intelligence Unit*. The authors of our books are acknowledged leaders in their fields and the topics are unique. Almost without exception, no other similar books exist on these topics.

Our goal is to publish books in important and rapidly changing areas of bioscience and the environment for sophisticated researchers and clinicians. To achieve this goal, we have accelerated our publishing program to conform to the fast pace in which information grows in the biosciences. Most of our books are published within 90 to 120 days of receipt of the manuscript. We would like to thank our readers for their continuing interest and welcome any comments or suggestions they may have for future books.

Shyamali Ghosh
Publications Director
R.G. Landes Company

DEDICATION

This book is dedicated to the memory of Frank Beerkenbosch—one of the major contributors to cytokine neurobiology, an outstanding scientist and a valued friend who is sadly missed.

CONTENTS

EDITOR

Nancy J. Rothwell
School of Biological Sciences
University of Manchester
Manchester, United Kingdom
Chapters 1,8

CONTRIBUTORS

Arnaud Aubert
INRA-INSERM
Bordeaux Cedex, France
Chapter 7

A. Bendele
Department of Inflammation
Amgen Inc.
Boulder, Colorado, U.S.A.
Chapter 9

Rose-Marie Bluthé
INRA-INSERM
Bordeaux Cedex, France
Chapter 7

Jean-Luc Bret-Dibat
INRA-INSERM
Bordeaux Cedex, France
Chapter 7

Carole A. Conn
Institute for Basic and Applied
  Medical Research
The Lovelace Institutes
Albuquerque, New Mexico, U.S.A.
Chapter 5

Robert Dantzer
INRA-INSERM
Bordeaux Cedex, France
Chapter 7

A. J. Davis
Sandoz Institute for Medical Research
London, U.K.
Chapter 10

Errol B. De Souza
Neurocrine Biosciences, Inc.
San Diego, California, U.S.A.
Chapter 3

N. Fischer
Department of Inflammation
Amgen Inc.
Boulder, Colorado, U.S.A.
Chapter 9

Philip W. Gold
Clinical Neuroendocrinology Branch
National Institute of Mental Health
National Institutes of Health
Bethesda, Maryland, U.S.A.
Chapter 2

Glyn Goodall
INRA-INSERM
Bordeaux Cedex, France
Chapter 7

Emmanuelle Goujon
INRA-INSERM
Bordeaux Cedex, France
Chapter 7

Kozo Hashimoto
Second Department of Internal
    Medicine
Kochi Medical School
Kohasu, Okoh-cho, Nankoku, Japan
Chapter 3

Keith W. Kelley
University of Illinois
Department of Animal Sciences
Urbana, Illinois, U.S.A.
Chapter 7

Stephen Kent
INRA-INSERM
Bordeaux Cedex, France
Chapter 7

Matthew J. Kluger
Institute for Basic and Applied
    Medical Research
The Lovelace Institutes
Albuquerque, New Mexico, U.S.A.
Chapter 5

Wieslaw Kozak
Institute for Basic and Applied
    Medical Research
The Lovelace Institutes
Albuquerque, New Mexico, U.S.A.
Chapter 5

James M. Krueger
Department of Physiology
    and Biophysics
University of Tennessee, Memphis
Memphis, Tennessee, U.S.A.
Chapter 4

Sophie Layé
INRA-INSERM
Bordeaux Cedex, France
Chapter 7

Lisa R. Leon
Institute for Basic and Applied
    Medical Research
The Lovelace Institutes
Albuquerque, New Mexico, U.S.A.
Chapter 5

Julio Licinio
Clinical Neuroendocrinology Branch
National Institute of Mental Health
National Institutes of Health
Bethesda, Maryland, U.S.A.
Chapter 2

Changlu Liu
Neurocrine Biosciences, Inc.
San Diego, California, U.S.A.
Chapter 3

D. Martin
Department of Inflammation
Amgen Inc.
Boulder, Colorado, U.S.A.
Chapter 9

G. Miller
Department of Inflammation
Amgen Inc.
Boulder, Colorado, U.S.A.
Chapter 9

Patricia Parnet
INRA-INSERM
Bordeaux Cedex, France
Chapter 7

M.N. Perkins
Sandoz Institute for Medical Research
London, U.K.
Chapter 10

J.K. Relton
Department of Inflammation
Amgen Inc.
Boulder, Colorado, U.S.A.
Chapter 9

Catherine Rivier
The Clayton Foundation Laboratories
   for Peptide Biology
The Salk Institute
La Jolla, California, U.S.A.
Chapter 6

D. Russell
Department of Inflammation
Amgen Inc.
Boulder, Colorado, U.S.A.
Chapter 9

Amer Al-Shekhlee
Clinical Neuroendocrinology Branch
National Institute of Mental Health
National Institutes of Health
Bethesda, Maryland, U.S.A.
Chapter 2

Dariusz Soszynski
Institute for Basic and Applied
   Medical Research
The Lovelace Institutes
Albuquerque, New Mexico, U.S.A.
Chapter 5

Toshihiro Takao
Second Department of Internal
   Medicine
Kochi Medical School
Kohasu, Okoh-cho, Nankoku, Japan
Chapter 3

Andrew V. Turnbull
The Clayton Foundation Laboratories
   for Peptide Biology
The Salk Institute
La Jolla, California, U.S.A.
Chapter 6

Ma-Li Wong
Clinical Neuroendocrinology Branch
National Institute of Mental Health
National Institutes of Health
Bethesda, Maryland, U.S.A.
Chapter 2

CHAPTER 1

INTRODUCTION

Nancy J. Rothwell

Cytokine neurobiology is now a hot topic! Only a few years ago most neuroscientists were only dimly aware of cytokines and knew little about their function or biological importance. Cytokines, now a huge collection of polypeptides with diverse activities, were until quite recently, studied by those interested in the immune system, inflammation, cancer or infection in peripheral tissues, and did not feature in neuroscience. For example, less than five years ago virtually no reference was made to cytokines in any of the numerous abstracts at the American Society for Neuroscience annual meeting.

This situation has now changed dramatically. In an article in early 1995 (Hopkins and Rothwell), we reported an exponential increase in articles on cytokines and the nervous system within the previous year and it seems that this publication frenzy is continuing. There are several reasons for such interest in cytokines and the nervous system. Firstly, the field of neuroimmunology (or psychoneuroimmunology/psychoneuroendocrinology) has developed considerably in the past five years. Thus the importance of interactions between the nervous, immune and endocrine systems in responses to disease, injury and stress is now recognized. These bidirectional communications have been mirrored by active dialogue (and even collaboration) between neuroscientists and immunologists. Cytokines form a critical part of neuroimmune interactions.

Secondly, it is now accepted that although the brain does not exhibit a classical immune response it does possess many of the mechanisms and molecules (notably cytokines and their receptors), which would normally be associated with the immune system. These processes are involved in many neurological diseases. Indeed, a major factor contributing to covert interest in cytokines and the nervous system is the emergence of their involvement in many neurological diseases and the potential they offer for new therapeutic approaches.

Having justified the need and timeliness of a book on cytokines and the nervous system, a much easier task is to justify the contributors.

Cytokines in the Nervous System, edited by Nancy J. Rothwell. © 1996 R.G. Landes Company.

All are international experts who have contributed significantly to the field and are at the forefront of cytokine neurobiology. This field is now so vast that it is inevitable there will be omissions in a short book, but the chapters prescribed cover the major areas of interest and current research on cytokine neurobiology. It is predicted that within a few years each chapter could itself form the subject of a book.

===== CHAPTER 2 =====

CYTOKINES IN THE BRAIN

Ma-Li Wong, Amer Al-Shekhlee, Philip W. Gold
and Julio Licinio

INTRODUCTION

Several cytokines and their receptors have been identified in the brain and in cerebrospinal fluid. Historically, the stimulatory action of interleukin-1 (IL-1) on the hypothalamic-pituitary-adrenal axis (HPA),[1] and the unexpected localization of IL-1β immunoreactivity neurons in the human hypothalamus,[2] have called attention to the role of cytokines in the brain. Since then, cytokines have been shown to play a key role in mediating communication between endocrine and immune systems.

During states of infection and inflammation, systemic cytokines cause clear biological effects mediated through the central nervous system, (CNS), such as fever, anorexia, and activation of the hypothalamic-pituitary-adrenocortical axis (HPA). The stimulatory action of IL-1 on the HPA has been extensively studied; IL-1 has been reported to enhance plasma level of ACTH by its action via the hypothalamus, where corticotropin releasing hormone (CRH) is synthesized and released into the portal veins. Other polypeptide cytokines, such as tumor necrosis factor alpha (TNF-α), interferon, and interleukin-6 (IL-6), have also been shown to alter neuroendocrine functions either via the central nervous system or through a direct action on endocrine cells.

Therapeutic uses of cytokines, and prolonged administration of interferons, interleukins and tumor necrosis factor are described to be accompanied by a range of toxic effects that vary from mild flu-like symptoms to CNS toxicity.[3-8] Cytokines can act in the brain through one or more of the following mechanisms: (1) disruption of the brain-blood-barrier (BBB); (2) penetration in the brain through circumventricular organs (these sites in the brain have capillaries with open junctions and abundant fenestrations); or (3) *de novo* synthesis in the CNS. It is still not known which of these mechanisms are involved in specific pathophysiological situations. Additionally, a crucial element for understanding the actions of cytokines in the CNS still needs completion, namely the mapping of cytokines and their receptors in the brain.

The localization and functional characterization of cytokines and their receptors in the brain lags behind the rapidly growing field of molecular characterization of cytokines and cytokines receptors in peripheral tissues, in large part due to the delay in the molecular characterization of the rat genes. While immunologists work with either human or mouse tissue, most of the classical neuroanatomical and neuroendocrinological studies have used the rat as an experimental animal. Though it is likely that researchers will have to focus in the mouse as an experimental animal, particularly because of the blooming field of transgenic animals, presently most of the understanding of functional neuroanatomy derives from studies done in rats.

We will summarize the actions and localization of cytokines in the CNS. A general summary of the possible toxic effects of cytokines by CNS anatomical sites can be found on Table 2.1,[8] the specific actions of each individual cytokine is somewhat obscured by the fact that many of their activities overlap and because cytokines regulate one another to form a cytokine network.[9] The hallmarks of cytokine biology are pleiotropism, redundancy and feedback.

INTERLEUKIN-1

IL-1, a pleiotropic cytokine, with multiple biological actions in various organs, including the central nervous system,[10] has been the focus of many studies addressing the actions of cytokines in the brain. The IL-1 system consists of two receptor types, the IL-1 type I (IL-1RI) and type II (IL-1RII) receptors, and three peptides: IL-1α, IL-1β and IL-1 receptor antagonist (IL-1ra), which is a unique naturally occurring antagonist. IL-1 and IL-1β activate the receptor, resulting in biological action, while IL-1ra completely blocks the receptor, preventing biological action.[11,12] IL-1β is converted from the pro IL-1β form to biologically active IL-1β by interleukin converting enzyme (ICE).[13,14] CNS actions of IL-1 include: regulation of temperature, food intake and neuroendocrine function.[15]

The actions of IL-1 are mediated by IL-1 receptors. IL-1RI has been identified in T cells, fibroblasts and other cell types.[16] The IL-1RII has been identified in B cells and macrophages.[17] While IL-1RI has a long cytoplasmic domain and is fully functional for signal transduction,[18] IL-1RII has a short cytoplasmic domain and it seems that it cannot mediate biological activity.[19]

IL-1 has been found in brain tissue: by detection of protein[20] and mRNA[21] IL-1β has been described in hypothalamic structures relevant for HPA, such as the paraventricular nucleus (PVN), and arcuate nucleus. It has been established that brain microglia expresses IL-1 through double labeling studies using immunohistochemistry and *in situ* hybridization;[21]

Table 2.1. *Toxic effects of cytokines in the CNS (modified from ref. 92)*

Anatomical areas	Biochemical and functional changes	Symptoms
CNS microvessel wall Endothelial cells Pericytes Astrocytes Neurons	Release of free radicals, PGs, etc. Modification of cations and H_2O distribution. Cellular edema. Alteration of neurotransmitter function.	confusion, visuospatial disorientation, headache, thought blocking, hallucinations, memory impairment, coma
Brain stem reticular formation	Effect of sleep-inducing substances: PGD_2, factor S, serotonin, IL-1	Hyersomnia, lethargy
Rostro-medical hypothalamus	Local release of IL-1 and/or PGE_2 Altered thermoregulation	Fever
Ventro medial hypothalamus activity	Increased cholecystokinin-like	Disregulation of appetite, anorexia

Table continues on next page.

Table 2.1. *(continued)*

Anatomical areas		Biochemical and functional changes	Symptoms
Circumventricular organs	Area postrema	Hemetic zone	Nausea and/or vomiting
	Organum vasculosum of the lamina terminalis	Disregulation of blood pressure	hypotension
	Hypothalamic neuro-secretory centers	Altered synthesis and release of hypothalamic releasing hormones. endorphins and neurotransmitters. Increased release of ACTH. Decrease of other pituitary hormones?	Analgesia, catatonia, depressed mood, self-pity, anxiety, impotence cortisol increase, SIADH*, hypothyroidism hypotestosteronemia
Neuro-muscular	Motor end-plate, skeletal muscle fiber	Changes in end-plate activity, decrease in muscle transmembrane potential difference, increased PGs release	Weakness, fatigue, asthenia, myalgias, paresthesiae
Gastrointestinal system	Endocrine and paracrine Cells (APUD cells). Acetylcholine-serotonin-, histamine-, PGs-releasing cells	Altered synthesis, release and metabolism of vasoactive intestinal polypeptide (VIP), gastric inhibitory polypeptide (GIP), cholecystokinin neurotensin, motilin, bombesin and neurotransmitters	Diarrheal syndrome

* Serum of inappropriate ADH secretion.

the confirmation that neuronal cells express IL-1 in several nuclei structures awaits carefully done double labeling studies. While the pathophysiological role of IL-1 is clearly shown in several paradigms, the understanding of constitutive actions of IL-1 in the brain is limited. Of note is the fact that a few tissues express IL-1 constitutively, for example, skin, amniotic fluid, sweat, and urine contain significant amounts of IL-1; macrophages and most other cell types, including astrocytes[22-26] produce IL-1 in response to external stimuli. In baseline conditions, in stress- and pathogen-free rats, mRNA for IL-1β and ICE have been localized in scattered vessels in the brain.[27]

IL-1ra, which is biologically inactive, but competes for binding of IL-1 receptors and is a competitive inhibitor of IL-1α and IL-1β, has also been reported to be present in the brain,[12] in areas relevant to HPA regulation, such as hippocampus and hypothalamus. Studies on the effects of exogenous IL-1ra have indicated that the blockage of endogenous IL-1 actions dramatically reduces brain damage caused by acute ischemic,[28] mechanical injury[29] and excitatory amino acids infusion.

IL-1 receptor type I (IL-1RI) gene expression and binding have been identified in the brain.[30-35] IL-1RI mRNA is present in high levels in endothelial cells of post-capillary venules, in glial cells surrounding arterioles, and in the choroid plexus. Though IL-1RI has been described in some neuronal cells (in the cerebellum and in the hippocampus), its absence in hypothalamic nuclei, especially the PVN, has been unexpected. Investigators have been trying to explain the hypothalamic actions of exogenous IL-1 in the PVN, speculating about the existence of other IL-1 receptors. It is unlikely that a new IL-1 receptor is present in the PVN, because of negative binding of labeled IL-1 in the PVN.[36] On the other hand, it is possible that IL-1β expressed in the PVN can act in receptors located in brain vasculature and have its biological activity tonically modulated through nitric oxide (NO) production (Fig. 2.1).[37,38] Alternatively, it is possible that some of the IL-1 synthesized in the hypothalamus might be transported to the median eminence and released in the portal circulation, having as its final target the receptors located in the pituitary gland.

Several lines of research suggest that endogenous IL-1 mediates neurodegeneration in the rat brain: IL-1 can also disturb the BBB,[39] elicit the release of arachidonic acid, NO and β amyloid precursor protein. IL-1b immunoreactivity has been shown to be increased in brain perivascular leukocytes and parenchymal microglial cells in brain tissue of patients with AIDS encephalitis.[40] ICE can also be involved in neuronal cell death.[41] Transgenic mice with a knock-out of either the IL-1b gene[42] or ICE gene[43,44] do not display any gross abnormal neurological defects and these mice will have to be carefully studied to ascertain any subtle CNS dysfunction.

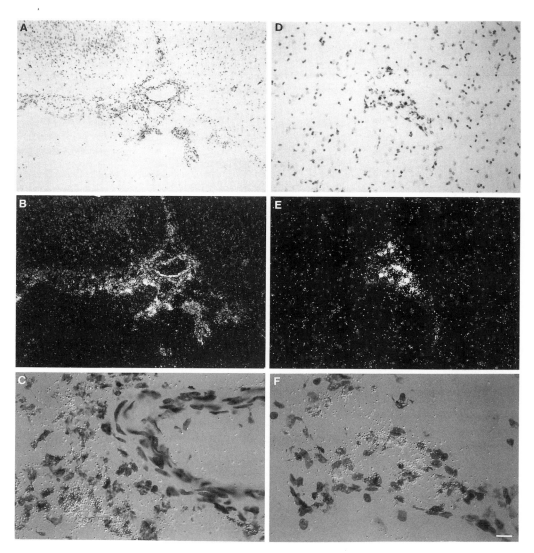

Fig. 2.1. Photomicrographs of iNOS mRNA in brain vessels. After treatment with LPS, cells strongly hybridized with iNOS mRNA probe can be seen in the area surrounding a small artery (A-C) and in the walls of a venule (D-F). The bright-field photomicrographs shown in the first row (A and D) correspond to the dark-field photomicrographs shown in the second row (B and E). High-magnification bright-field photomicrographs are shown in C and F. Silver grains overlying iNOS mRNA are shown as white dots in B and E, and in bright lighter gray dots over cells in C and F. Scale bar, 240 μm for A and B; 120 μm for D and E; and 15 μm for C and F.

TUMOR NECROSIS FACTOR (TNF)

TNF and IL-1 are structurally unrelated cytokines that bind to different cellular receptors, but yet their spectra of biologic actions overlap considerably. CNS actions of TNF-α, are similar to the actions of IL-1: TNF induces fever, sickness behavior, and directly induces the secretion of corticotropin-releasing hormone.

There are two forms of TNF: TNF-α and TNF-β, and they both bind to the same receptor on target cells, therefore they have the same biologic activities in spite of having only 28% similarity at the amino acid level. The genes encoding for both TNF-α and TNF-β are found within the Major Histocompatibility Complex (MHC) on chromosome 6. TNF-α is widely expressed in several cell types in the immune system, and in glial cells and astrocytes.[45,46] TNF-β can be produced by several cell types, including lymphocytes, astrocytes, lymphokine-activated killer cells and myeloma cells.[45,46] Inducers of TNF production include IL-1, IL-2, interferons, endotoxin and mitogens. Suppressors of TNF production include IL-4, IL-6, transforming growth factor β and dexamethasone.

Two types of high affinity receptors have been identified for TNF: TNF receptor type I (TNFRI) and TNF receptor type II (TNFRII).[47] TNFRII binds TNF with about 10-fold higher affinity than TNFRI. The two receptors are independently regulated and transduce distinct intracellular signals, and they are thought to elicit distinct responses: the type I receptor promotes cytotoxic activity and fibroblast proliferation, and the type II receptor promotes T lymphocyte proliferation.

TNF can be synthesized and released by astrocytes, microglia and some neurons,[24-26,48] and TNF can induce the proliferation of astrocytes.[49] TNF has been recently shown to influence neuronal progenitor cell proliferation and differentiation.[50-52] Detailed localization of TNF has been examined by immunoreactivity,[53] and by in situ hybridization[54] in mouse brain. In baseline conditions TNF immunoreactivity was primarily observed in neuronal structures, such as fibers and terminal fields;[53] colchicine treatment increased the intensity of immunoreactivity staining and allowed the visualization of neuronal cell bodies in hypothalamic nuclei: periventricular preoptic nucleus, paraventricular nucleus, ventromedial nucleus, dorsomedial nucleus. Specific groups of neurons were found to have immunreactivity to TNF-α in the pons and medulla. TNF-α mRNA[54] was observed in baseline conditions in regions that TNF-α immunoreactivity was observed, these areas included the periventricular preoptic and suprachiasmatic nuclei in the hypothalamus and in the preoptic portion of the bed nucleus of the stria terminalis and the ventral medulla. After systemic administration of lipopolysaccharide (LPS) intense hybridization signal was found predominantly in circumventricular organs (organum vasculosum laminae terminalis [OVLT], median eminence, area postrema), meninges, and arcuate nucleus of the hypothalamus. Interestingly, the OVLT is also a brain site that produces interleukin-1 β during fever.[55]

The expression of TNF receptors has been shown to be upregulated by several leukocyte-produced cytokines such as IL-1β, IFN-γ, TNF-α[56-58] in vitro. Modulation of TNF receptor expression by inflammatory cytokines may represent one level of control of TNF responsiveness. TNF receptors are widely distributed throughout most cells and tissues, including the brain.[59] In vitro studies have shown that human microglia, astrocytes, and oligodendrocytes express TNFRI and TNFRII.[60,61] TNF receptors[40] and TNF-α[62,63] have both been shown to be increased in brain tissue of AIDS patients. Increased TNF receptors immunoreactivity have also been found in patients with multiple sclerosis, chronic cerebral edema and radiation necrosis.

It has been speculated that TNF-α may function as a mediator of cerebral damage. Even though TNF may mediate damage to myelin and oligodendrocytes, it is still unclear whether this cytokine is neurotoxic or neuroprotective: TNF has been reported to be non-toxic for CNS neurons *in vitro*, to facilitate regeneration of injured axons,[64] and to protect cultured embryonic rat hippocampal, septal and cortical cells.[65] Despite the fact that TNF expression is increased in AIDS[62,63] and also in response to brain injury,[66] the actions of TNF in neurons are unknown.

INTERLEUKIN-6 (IL-6)

IL-6 is a pleiotropic cytokine that synergizes with IL-1 and TNF to co-stimulate immune responses. IL-6 major activities include inducing the acute-phase response in the liver, enhancing B cell replication, differentiation and immunoglobulin production. IL-6 can be produced by many cell types, such as T and B lymphocytes, monocytes, endothelial cells, and fibroblasts. The gene for IL-6 is located on human chromosome 7 and its expression may be induced by a variety of stimuli, including TNF and IL-1.

IL-6 receptor consists of 2 glycoprotein chains: IL-6Rα and IL-6Rβ. IL-6Rα is a low affinity chain and lacks a cytoplasmic domain. The complex IL-6 and IL-6Rα binds to the IL-6Rβ chain, which has a high affinity to IL-6 and transduces the intracellular signal. The IL-6Rβ chain is found to form part of other structurally unrelated cytokine receptors, such as IL-11, leukocyte inhibitory factor (LIF), oncostatin M (OSM), and ciliary neurotrophic factor (CNTF) and functions as a common signaling transducing portion in each of these receptors.

IL-6 has prominent effects on the CNS: these central effects include activation of the hypothalamic-pituitary-adrenal axis, reduction of food intake, induction of fever and neuronal growth.[9] Astrocytes and microglia can be stimulated in vitro to produce IL-6,[67] and high brain IL-6 levels have been demonstrated in patients with acute infection of the CNS in the absence of elevated blood IL-6 concentrations.[68] IL-6 has been described to promote survival of mesencephalic catecholaminergic and septal cholinergic neurons in vitro;[69] to induce neuronal differentiation of the rat pheochromocytoma cell line PC12,[70] and to significantly ameliorate the neurotoxic effects of N-methyl-D-aspartate (NMDA) on rat striatal cholinergic neurons. Localization of IL-6 gene expression has been described in the rat brain using oligonucleotide probes[71] and cRNA probes:[72] mRNA for IL-6 was found to be generally low in the brain[73] and was present in the hippocampal formation with highest signal in the dentate gyrus, habenular nucleus, piriform cortex, hypothalamus and striatum. Localization of IL-6R mRNA was described to have the same pattern of distribution of IL-6 mRNA.[71] The highly discrete density of signal in the granular cells of the dentate gyrus and the pyramidal cells of the hippocampus could suggest neuronal localization of the mRNAs. Double labeling experiments have not been done yet to confirm the identity of the cells that express IL-6 and IL-6 receptors in the central nervous system.

IL-6 expression is elevated in the cortex of brains from Alzheimer's disease patients,[74] and has been specifically co-localized with α2-microglobulin, an acute phase protein. While the role of the increased levels of IL-6 in the brain under pathological conditions are unclear, transgenic mice overexpressing IL-6 in glial cells[75] have a neurologic syndrome that is correlated with the levels of IL-6 expression in the brain. The striking proliferative angiopathy in the brain, which caused considerable distortion of surrounding tissue in the transgenic mice overexpressing IL-6, implicates this cytokine in angiogenesis.[75-77] The alterations found in the CNS of transgenic mice overexpressing IL-6, such as loss of neuronal subpopulations and significant change in neuronal morphology, overlap findings in autoimmune, infectious, degenerative, and trauma-induced CNS diseases.

TRANSFORMING GROWTH FACTOR (TGF)

TGFβ was discovered as a growth factor for fibroblasts that promote wound healing. TGFβ has antiproliferative effects; it also suppresses the production of most cytokines that have chemoattractant activity for leukocytes and fibroblasts, and of cytokines that are produced by monocytes or macrophages. It reduces the cellular expression of class II MHC proteins and IL-1 receptors. TGFβ has also been demonstrated to control differentiation and morphogenesis in embryonic development. TGFβ is produced by many cell types, including activated macrophages and T lymphocytes. It is now apparent that there are at least three distinct isoforms of TGFβ: TGFβ1, TGFβ2 and TGFβ3 in mammals. These are encoded by different genes, but they all act in the same five types of high-affinity cell surface receptors. There are three types of TGFβ receptors: type I and type II receptors have a high affinity for TGFβ and are known to transduce signals; type III TGFβ receptor, also known as betaglycan, binds to TGFβ with lower affinity.[78]

TGFβ1 mRNA has been either detected in the brain in very low levels,[79,80] or found to be undetectable[81] in basal conditions, while TGFβ1 immunoreactivity has been localized in the meninges and in the choroid plexus in basal conditions. TGFβ1 mRNA levels have been reported to be increased in hippocampus after ischemia[79,81,82] injury,[80] and in response to deafferentiation and kainic acid-induced neurodegeneration.[83]

Increased levels of TGFβ1 protein have also been found in brains of patients with neurologic disease such as AIDS dementia[84,85] and Alzheimer's disease. In AIDS,[84,85] and in lesioned rat brains, expression of TGFβ1 was confined to astrocytes and microglia, and in Alzheimer's disease TGFβ1 immunoreactivity was found in amyloid plaques.

In vivo and/or in vitro studies of the effects of TGFβ1 actions, showed that TGFβ1 increases GFAP mRNA, and NGF mRNA.[80] TGFβ1 can autoregulate its production through the induction of TGFβ1 gene transcription. In vitro studies showed that TGFβ1 is chemotactic for astrocytes, inhibits their proliferation,[86] and potentiates the mitogenic effect of bFGF.[87]

While TGFβ receptor type II mRNA has been recently described to be strictly localized to the ventral midline cells of the hindbrain and spinal cord in murine embryonic development,[88] it is clear that the field needs further detailed studies on the localization of the TGFβ receptors in view of the critical role TGFβ1 might have in the brain response to neurogeneration.

A possible role of TGFβ1 during responses to brain lesions is the regulation of astrocytic functions and the present hypothesis is that microglial TGFβ1 has a role in mediating early responses to terminal degeneration and/or neurodegeneration. While it remains unclear whether the overproduction of TGFβ1 in neurological conditions has primary beneficial or detrimental effects in the CNS, transgenic mice overexpressing TGFβ1 in astrocytes have increased CNS production of extracellular matrix components that can induce neurological disease.

OTHER INTERLEUKINS AND INTERFERONS (IFNs)

IL-3 was one of the first cytokines to be discovered in the brain.[89] Recently, Chiang et al[90] generated transgenic mice with a murine glial fibrillary acidic protein fusion gene. Those animals overexpressed low levels of IL-3 in astrocytes and developed from 5 months of age, a progressive motor disorder characterized at onset by impaired rota-rod performance. In symptomatic transgenic mice, multi-focal, plaque-like white matter lesions were present in cerebellum and brain stem. Lesions showed extensive primary demyelination and remyelination in association with the accumulation of large numbers of proliferating and activated foamy macrophage/microglial cells. Many of these cells also contained intracisternal crystalline pole-like inclusions similar to those seen in human patients with multiple sclerosis. Those authors conclude that the central IL-3 overexpression transgenic model exhibits many of the features of human inflammatory demyelinating diseases including multiple sclerosis and HIV leukoencephalopathy.

Interferons (IFNs) are a large family of secretory proteins that have antiviral activity and also the ability to inhibit proliferation of vertebrate cells and to modulate immune responses. Interferons induce an antiviral state within the host cell that makes it hospitable to viral replication. Moreover, because many different types of proteins can induce an antiviral state in vertebrate cells, their molecular and biological properties differ widely. IFN-α has been shown to be effective in the treatment of several virus pathologies, such as chronic active hepatitis B; life threatening, virus-induced juvenile laryngeal papillomatosis, and herpes zoster and simplex infections. IFN has also been used as an antitumor agent.[91] Results of preliminary trials of the interferons as antiviral and antitumor agents have been encouraging and occasionally dramatic, but chronic administration of IFNs can cause a range of side effects. The CNS side effects are the most problematic and can be dose limiting; such side effects include nausea, fatigue, confusion, coma, paresthesia, psychomotor slowing, speech stoppage, thought blocking and visual disorientation.[92]

CONCLUDING REMARKS

Discrepancies of findings in localization studies, particularly those examining basal levels of cytokines or their receptors in the brain, might be attributed to one of several explanations: (1) differences in paradigms, especially in the time course of experiments; (2) species differences in baseline levels or in the response to stimulation; (3) differences in the baseline state of the animals; (4) differences in probes or antibodies used; however, now the availability of species-specific probes ought to make conclusive studies possible. The baseline state of the animals might account for the discrepancies on the constitutive levels of cytokines found in the brain, because rapid induction of the mRNAs encoding cytokine and their receptors have been reported to occur as a result of stress, inflammation, or infection.

Over 40 cytokines have been identified. Most of them (and their receptors) are expressed in the brain and affect key functions of the central nervous system. Each cytokine has its own specific pattern of gene expression, and specific temporal and spatial responses to various stimuli. Each brain cytokine has multiple actions (pleiotropiosm), and the actions of various cytokines overlap (redundancy). Moreover, various stimuli lead the brain to synthesize a complex cascade of primary and secondary cytokines, with multiple feedback loops. Even though considerable progress has been made in the field, additional studies are needed to elucidate at the descriptive level the patterns of cytokine gene expression in different brain regions after various stimuli. On a mechanistic level, we still need to address the key question: How does cytokine-mediated signal transduction occur in the brain? Finally, simple but crucial questions in the field remain unanswered: What is the communication mechanism between peripheral cytokines and brain cytokines? Do cytokines have a role in the normal functioning of the brain or are they elicited only during pathological processes? Do cytokines have a primary role in the pathogenesis of neuropsychiatric disorders? Can strategies that affect cytokine bioactivity be used in the treatment of disorders of the brain? Intensive work in this exciting field by an increasing number of investigators worldwide is bringing us closer to answers for these important questions that are relevant not only to specialists in neuroscience and neuroimmunology but to all who practice medicine.

REFERENCES

1. Sapolsky R, Rivier C, Yamamoto G, Plotsky P, Vale W. Interleukin-1 stimulates the secretion of hypothalamic corticotropin releasing factor. Science 1987; 238:522-524.
2. Breder CD, Dinarello CA, Saper CB. Interleukin-1 immunoreactive innervation of the human hypothalamus. Science 1988; 240:321-324.
3. Rothwell NJ. Eicosanoids, thermogenesis and thermoregulation. Prostaglandins Leukot Essent Fatty Acids 1992; 46(1):1-7.
4. Cooper AL, Horan MA, Little RA, Rothwell NJ. Metabolic and febrile responses to typhoid vaccine in humans: effect of beta-adrenergic blockade. J Appl Physiol 1992; 72(6):2322-8.
5. Avitsur R, Donchin O, Barak O, Cohen E, Yirmiya R. Behavioral effects of interleukin 1 beta modulation by gender, estrus cycle, and progesterone. Brain, Behavior & Immunity 1995; 9(3):234-241.
6. Strijbos PJ, Horan MA, Carey F, Rothwell NJ. Impaired febrile responses of aging mice are mediated by endogenous lipocortin-1 (annexin-1). Am J Physiol 1993; 265(2 Pt 1):e289-97.
7. Rothwell NJ, Relton JK. Involvement of cytokines in acute neurodegeneration in the CNS. Neurosci Biobehav Rev 1993; 17(2):217-27.
8. Dantzer R. [Neurotropic effects of cytokines: at the limits of immunology and neurobiology (editorial)]. Pathol Biol (Paris) 1994; 42(9):826-9.
9. Akira S, Hirano T, Taga T, Kishimoto T. Biology of multifunctional cytokines: IL 6 and related molecules (IL 1 and TNF). FASEB J 1990; 4(11):2860-7.
10. Dinarello CA, Wolff SM. The role of interleukin-1 in disease. N Engl J Med 1993; 328:106-113.
11. Lee S, Rivier C. Prenatal alcohol exposure alters the hypothalamic-pituitary-adrenal axis response of immature offspring to interleukin-1: is nitric oxide involved? Alcohol Clin Exp Res 1994; 18(5):1242-7.
12. Licinio J, Wong ML, Gold PW. Localization of interleukin-1 receptor antagonist mRNA in rat brain. Endocrinology 1991; 129(1):562-4.
13. Thornberry NA, Bull HG, Calaycay JR et al. A novel heterodimeric cys-

teine protease is required for interleukin-1 beta processing in monocytes. Nature 1992; 356:768-774.

14. Cerretti DP, Kozlosky CJ, Mosley B et al. Molecular cloning of the interleukin-1 beta converting enzyme. Science 1992; 256:97-100.

15. Rothwell NJ, Hopkins SJ. Cytokines and the nervous system II: Actions and mechanisms of action. Trends Neurosci 1995; 18(3):130-136.

16. Sims JE, March CJ, Cosman D et al. cDNA expression cloning of the IL-1 receptor, a member of the immunoglobulin superfamily. Science 1988; 241:585-589.

17. McMahan CJ, Slack JL, Mosley B et al. A novel IL-1 receptor, cloned from B-cells by mammalian expression, is expressed in many cell types. EMBO J 1991; 10:2821-2832.

18. Ye K, Dinarello CA, Clark BD. Identification of the promoter region of human interleukin 1 type I receptor gene: multiple initiation sites, high G+C content, and constitutive expression. Proc Natl Acad Sci USA 1993; 90(6):2295-9.

19. Colotta F, Re F, Muzio M et al. Interleukin-1 type II receptor: a decoy target for IL-1 that is regulated by IL-4. Science 1993; 261:472-475.

20. Bluthe RM, Dantzer R, Kelley KW. Interleukin-1 mediates behavioural but not metabolic effects of tumor necrosis factor alpha in mice. Eur J Pharmacol 1991; 209(3):281-3.

21. Buttini M, Boddeke H. Peripheral lipopolysaccharide stimulation induces interleukin-1 beta messenger RNA in rat brain microglial cells. Neuroscience 1995; 65(2):523-30.

22. Malipiero UV, Frei K, Fontana A. Production of hemopoietic colony-stimulating factors by astrocytes. J Immunol 1990; 144(10):3816-21.

23. Fontana A, Hengartner H, Weber E, Fehr K, Grob PJ, Cohen G. Interleukin 1 activity in the synovial fluid of patients with rheumatoid arthritis. Rheumatol Int 1982; 2(2):49-53.

24. Chung IY, Benveniste EN. Tumor necrosis factor-alpha production by astrocytes. Induction by lipopolysaccharide, IFN-gamma, and IL-1 beta. J Immunol 1990; 144(8):2999-3007.

25. Lieberman AP, Pitha PM, Shin HS, Shin ML. Production of tumor necrosis factor and other cytokines by astrocytes stimulated with lipopolysaccharide or a neurotropic virus. Proc Natl Acad Sci USA 1989; 86(16):6348-52.

26. Sawada M, Kondo N, Suzumura A, Marunouchi T. Production of tumor necrosis factor-alpha by microglia and astrocytes in culture. Brain Res 1989; 491(2):394-7.

27. Wong M-L, Bongiorno PB, Gold PW, Licinio J. Localization of interleukin 1β (IL-1β) converting enzyme mRNA in rat brain vasculature: evidence that the genes encoding the IL-1 system are constitutively expressed in brain blood vessels. Pathophysiological Implications. Neuroimmunomodulation 1995; 2:141-148.

28. Relton JK, Rothwel NJ. Interleukin-1 receptor antagonist inhibits ischaemic and excitotoxic neuronal damage in the rat. Brain Res Bull 1992;

29:243-246.

29. Toulmond S, Rothwell NJ. Interleukin-1 receptor antagonist inhibits neuronal damage caused by fluid percussion injury in the rat. Brain Res 1995; 671(2):261-6.

30. Wong M-L, Licinio J. Localization of interleukin 1 type I receptor mRNA in rat brain. Neuroimmunomodulation 1994; 1:110-115.

31. Rivest S, Rivier C. Interleukin-1 beta inhibits the endogenous expression of the early gene c-fos located within the nucleus of LH-RH neurons and interfers with hypothalamic LH-RH release during proestrus in the rat. Brain Res 1993; 613(1):132-42.

32. Takao T, Tracey DE, Mitchell WM, De Souza EB. Interleukin-1 receptors in mouse brain: characterization and neuronal localization. Endocrinol 1990; 127:3070-3078.

33. Ericsson A, Liu C, Hart RP, Sawchenko PE. Type 1 interleukin-1 receptor in the rat brain: distribution, regulation, and relationship to sites of IL-1-induced cellular activation. J Comp Neurol 1995; 361(4):681-98.

34. Cunningham ETJ, Wada E, Carter DB, Tracey DE, Battey JF, De Souza EB. In situ histochemical localization of type I interleukin-1 receptor messenger RNA in the central nervous system, pituitary, and adrenal gland of the mouse. J Neurosci 1992; 12:1101-1114.

35. Yabuuchi K, Minami M, Katsumata S, Satoh M. Localization of type I interleukin-1 receptor mRNA in the rat brain. Brain Res Mol Brain Res 1994; 27(1):27-36.

36. Ban E, Milon G, Prudhomme N, Fillion G, Haour F. Receptors for interleukin-1 (α and β) in mouse brain: mapping and neuronal localization in hippocampus. Neurosci 1991; 43:21-30.

37. Wong M-L, Bongiorno PB, Al-Shekhlee A et al. IL-1β, IL-1 Receptor Type I, and iNOS gene expression in rat brain vasculature and perivascular areas. NeuroReport 1996; in press.

38. Wong M-L, Rettori V, Al-Shekhlee A et al. Inducible nitric oxide synthase gene expression in the brain during systemic inflammation. Nature Med 1996; 2:581-584.

39. Quagliarello VJ, Wispelwey B, Long WJ, Scheld WM. Recombinant human interleukin-1 induces meningitis and blood-brain barrier injury in the rat. Characterization and comparison with tumor necrosis factor. J Clin Invest 1991; 87(4):1360-6.

40. Sippy BD, Hofman FM, Wallach D, Hinton DR. Increased expression of tumor necrosis factor-alpha receptors in the brains of patients with AIDS. J Acquir Immune Defic Syndr Hum Retrovirol 1995; 10(5):511-21.

41. Gagliardini V, Fernandez PA, Lee RK et al. Prevention of vertebrate neuronal death by the crmA gene. Science 1994; 263:826-828.

42. Zheng H, Fletcher D, Kozak W et al. Resistance to fever induction and impaired acute-phase response in interleukin-1 beta-deficient mice. Immunity 1995; 3(1):9-19.

43. Li P, Allen H, Banerjee S et al. Mice deficient in IL-1β converting enzyme are defective in production of mature IL-1β and resistant to

endototoxic shock. Cell 1995; 80:401-411.

44. Kuida K, Lippke JA, Ku G, et al. Altered cytokine export and apoptosis in mice deficient in interleukin-1β converting enzyme. Science 1995; 267:2000-2003.

45. Aggarwal BB. Comparative analysis of the structure and function of TNF-alpha and TNF-beta. Immunol Ser 1992; 56(61):61-78.

46. Turetskaya RL, Fashena SJ, Paul NL, Ruddle NH. Genomic structure, induction, and production of TNF-beta. Immunol Ser 1992; 56(35):35-60.

47. Tartaglia LA, Weber RF, Figari IS, Reynolds C, Palladino MJ, Goeddel DV. The two different receptors for tumor necrosis factor mediate distinct cellular responses. Proc Natl Acad Sci USA 1991; 88(20):9292-6.

48. Gendron RL, Nestel FP, Lapp WS, Baines MG. Expression of tumor necrosis factor alpha in the developing nervous system. Int J Neurosci 1991; 60(1-2):129-36.

49. Selmaj KW, Farooq M, Norton WT, Raine CS, Brosnan CF. Proliferation of astrocytes in vitro in response to cytokines. A primary role for tumor necrosis factor. J Immunol 1990; 144(1):129-35.

50. Bazan JF. Neuropoietic cytokines in the hematopoietic fold. Neuron 1991; 7(2):197-208.

51. Merrill JE. Tumor necrosis factor alpha, interleukin 1 and related cytokines in brain development: normal and pathological. Dev Neurosci 1992; 14(1):1-10.

52. Mehler MF, Rozental R, Dougherty M, Spray DC, Kessler JA. Cytokine regulation of neuronal differentiation of hippocampal progenitor cells. Nature 1993; 362(6415):62-5.

53. Breder CD, Tsujimoto M, Terano Y, Scott DW, Saper CB. Distribution and characterization of tumor necrosis factor-alpha-like immunoreactivity in the murine central nervous system. J Comp Neurol 1993; 337(4):543-67.

54. Breder CD, Hazuka C, Ghayur T et al. Regional induction of tumor necrosis factor alpha expression in the mouse brain after systemic lipopolysaccharide administration. Proc Natl Acad Sci USA 1994; 91:11393-97.

55. Nakamori T, Morimoto A, Yamaguchi K, Watanabe T, Long NC, Murakami N. Organum vasculosum laminae terminalis (OVLT) is a brain site to produce interleukin-1 beta during fever. Brain Res 1993; 618(1):155-9.

56. Pandita R, Pocsik E, Aggarwal BB. Interferon-gamma induces cell surface expression for both types of tumor necrosis factor receptors. FEBS Lett 1992; 312(1):87-90.

57. Winzen R, Wallach D, Kemper O, Resch K, Holtmann H. Selective upregulation of the 75-kDa tumor necrosis factor (TNF) receptor and its mRNA by TNF and IL-1. J Immunol 1993; 150(10):4346-53.

58. Trefzer U, Brockhaus M, Lotscher H et al. The 55-kD tumor necrosis factor receptor on human keratinocytes is regulated by tumor necrosis factor-alpha and by ultraviolet B radiation. J Clin Invest 1993;

92(1):462-70.

59. Kinouchi K, Brown G, Pasternak G, Donner DB. Identification and characterization of receptors for tumor necrosis factor-alpha in the brain. Biochem Biophys Res Commun 1991; 181(3):1532-8.

60. Wilt SG, Milward E, Zhou JM et al. In vitro evidence for a dual role of tumor necrosis factor-alpha in human immunodeficiency virus type 1 encephalopathy [see comments]. Ann Neurol 1995; 37(3):381-94.

61. Tada M, Diserens AC, Desbaillets I, de TN. Analysis of cytokine receptor messenger RNA expression in human glioblastoma cells and normal astrocytes by reverse-transcription polymerase chain reaction. J Neurosurg 1994; 80(6):1063-73.

62. Merrill JE, Chen IS. HIV-1, macrophages, glial cells, and cytokines in AIDS nervous system disease. FASEB J 1991; 5(10):2391-7.

63. Merrill JE, Koyanagi Y, Zack J, Thomas L, Martin F, Chen IS. Induction of interleukin-1 and tumor necrosis factor alpha in brain cultures by human immunodeficiency virus type 1. J Virol 1992; 66(4):2217-25.

64. Schwartz M, Solomon A, Lavie V, Ben BS, Belkin M, Cohen A. Tumor necrosis factor facilitates regeneration of injured central nervous system axons. Brain Res 1991; 545(1-2):334-8.

65. Cheng B, Christakos S, Mattson MP. Tumor necrosis factors protect neurons against metabolic-excitotoxic insults and promote maintenance of calcium homeostasis. Neuron 1994; 12(1):139-53.

66. Tchelingerian JL, Quinonero J, Booss J, Jacque C. Localization of TNF alpha and IL-1 alpha immunoreactivities in striatal neurons after surgical injury to the hippocampus. Neuron 1993; 10(2):213-24.

67. Frei K, Malipiero UV, Leist TP, Zinkernagel RM, Schwab ME, Fontana A. On the cellular source and function of interleukin 6 produced in the central nervous system in viral diseases. Eur J Immunol 1989; 19(4):689-94.

68. Houssiau FA, Bukasa K, Sindic CJM. Elevated levels of the 26K human hybridoma growth factor (interleukin 6) in cerebrospinal fluid of patients with acute infection of the central nervous system. Clin Exp Immunol 1988; 71:320-323.

69. Hama T, Kushima Y, Miyamoto M, Kubota M, Takei N, Hatanaka H. Interleukin-6 improves the survival of mesencephalic catecholaminergic and septal cholinergic neurons from postnatal, two-week-old rats in cultures. Neuroscience 1991; 40(2):445-52.

70. Satoh T, Endo M, Nakamura S, Kaziro Y. Analysis of guanine nucleotide bound to ras protein in PC12 cells. FEBS Lett 1988; 236(1):185-9.

71. Schobitz B, Holsboer F, Kikkert R, Sutanto W, De KE. Peripheral and central regulation of IL-6 gene expression in endotoxin-treated rats. Endocr Regul 1992; 26(3):103-9.

72. Gadient RA, Otten U. Interleukin-6 and interleukin-6 receptor mRNA expression in rat central nervous system. Ann NY Acad Sci 1995; 762(403):403-6.

73. Schobitz B, de Kloet ER, Sutanto W, Holsboer F. Cellular localization of

interleukin 6 mRNA and interleukin 6 receptor mRNA in rat brain. Eur J Neurosci 1993; 5:1426-1435.

74. Huell M, Strauss S, Volk B, Berger M, Bauer J. Interleukin-6 is present in early stages of plaque formation and is restricted to the brains of Alzheimer's disease patients. Acta Neuropathol (Berl) 1995; 89(6):544-51.

75. Campbell IL, Abraham CR, Masliah E et al. Neurologic disease induced in transgenic mice by cerebral overexpression of interleukin 6. Proc Natl Acad Sci USA 1993; 90(21):10061-5.

76. Motro B, Itin A, Sachs L, Keshet E. Pattern of interleukin 6 gene expression in vivo suggests a role for this cytokine in angiogenesis. Proc Natl Acad Sci USA 1990; 87(8):3092-6.

77. Rosen EM, Liu D, Setter E, Bhargava M, Goldberg ID. Interleukin-6 stimulates motility of vascular endothelium. Exs 1991; 59(194):194-205.

78. Massague J, Andres J, Attisano L et al. TGF-beta receptors. Mol Reprod Dev 1992; 32(2):99-104.

79. Klempt ND, Sirimanne E, Gunn AJ et al. Hypoxia-ischemia induces transforming growth factor beta 1 mRNA in the infant rat brain. Brain Res Mol Brain Res 1992; 13(1-2):93-101.

80. Lindholm D, Castren E, Kiefer R, Zafra F, Thoenen H. Transforming growth factor-beta 1 in the rat brain: increase after injury and inhibition of astrocyte proliferation. J Cell Biol 1992; 117(2):395-400.

81. Lehrmann E, Kiefer R, Finsen B, Diemer NH, Zimmer J, Hartung HP. Cytokines in cerebral ischemia: expression of transforming growth factor beta-1 (TGF-beta 1) mRNA in the postischemic adult rat hippocampus. Exp Neurol 1995; 131(1):114-23.

82. Logan A, Frautschy SA, Gonzalez AM, Sporn MB, Baird A. Enhanced expression of transforming growth factor beta 1 in the rat brain after a localized cerebral injury. Brain Res 1992; 587(2):216-25.

83. Morgan TE, Nichols NR, Pasinetti GM, Finch CE. TGF-beta 1 mRNA increases in macrophage/microglial cells of the hippocampus in response to deafferentation and kainic acid-induced neurodegeneration. Exp Neurol 1993; 120(2):291-301.

84. Vitkovic L, da Cunha A, Tyor WR. Cytokine expression and pathogenesis in AIDS brain. Res Publ Assoc Res Nerv Ment Dis 1994; 72:203-222.

85. Wahl SM, Allen JB, McCartney FN et al. Macrophage- and astrocyte-derived transforming growth factor beta as a mediator of central nervous system dysfunction in acquired immune deficiency syndrome. J Exp Med 1991; 173(4):981-91.

86. Morganti KM, Kossmann T, Brandes ME, Mergenhagen SE, Wahl SM. Autocrine and paracrine regulation of astrocyte function by transforming growth factor-beta. J Neuroimmunol 1992; 39(1-2):163-73.

87. Labourdette G, Janet T, Laeng P, Perraud F, Lawrence D, Pettmann B. Transforming growth factor type beta 1 modulates the effects of basic fibroblast growth factor on growth and phenotypic expression of rat astroblasts in vitro. J Cell Physiol 1990; 144(3):473-84.

88. Wang Y-Q, Sizeland A, Wang X-F, Sassoon D. Restricted expression of

type II TGFβ receptor in murine embryonic development suggests a central role in tissue modeling and CNS patterning. Mechanisms of Development 1995; 52:275-258.

89. Farrar WL, Vinocour M, Hill JM. In situ hybridization histochemistry localization of interleukin-3 mRNA in mouse brain. Blood 1989(73); 137-140.

90. Chiang CS, Powell HC, Gold LH, Samimi A, Campbell IL. Macrophage/microglial-mediated primary demyelination and motor disease induced by the central nervous system production of interleukin-3 in transgenic mice. J Clin Invest 1996; 97(6):1512-24.

91. Mannering GJ DL. The pharmacology and toxicology of interferons: an overview. Ann Rev Pharmacol Toxicol 1986; 26:455-515.

92. Bocci V. Central nervous system toxicity of interferons and other cytokines. J Biol Regul Homeost Agents 1988; 2(3):107-18.

CHAPTER 3

INTERLEUKIN-1 RECEPTORS IN THE NERVOUS SYSTEM

Changlu Liu, Toshihiro Takao, Kozo Hashimoto and Errol B. De Souza

INTRODUCTION

The cytokine interleukin-1 (IL-1) is a hormone-like polypeptide that performs many roles in inflammation and immunity.[1-3] Currently, two forms of IL-1 (IL-1α and IL-1β) and one IL-1 receptor antagonist (IL-1ra) have been characterized.[1] In addition to its immune effects, a role has been postulated for IL-1 as a neurotransmitter/neuromodulator/ growth factor in the central nervous system (CNS). IL-1 production has been reported in cultured brain astrocytes and microglia[4-6] and IL-1 has been detected in the brain following cerebral trauma[7,8] and endotoxin treatment.[9] IL-1-like activity is also present in the cerebrospinal fluid (CSF),[10,11] IL-1 mRNA is present in normal brain,[12,13] and immunohistochemical studies have identified neurons positive for IL-1β-like immunoreactivity in both hypothalamic[14,15] and extrahypothalamic[14] sites in human brain. Administration of IL-1 in brain produces a variety of effects, including induction of fever,[1-3] alteration of slow-wave sleep,[16,17] reduction of food intake,[18] induction of analgesia,[19] induction of acute-phase glycoprotein synthesis,[20,21] stimulation of thermogenesis[22,23] and reduction of peripheral cellular immune responses.[24] Central as well as peripheral administration of IL-1 has potent neuroendocrine actions including stimulation of the hypothalamic-pituitary-adrenocortical axis[25-27] and inhibition of the hypothalamic-pituitary-gonadal axis.[28] These central effects of IL-1 are presumably mediated through actions of the cytokine at specific high affinity receptors. In this chapter, we summarize some of the data from our recent molecular biological and radioligand binding studies to identify, characterize and localize IL-1 receptors in the central nervous system. In addition,

Cytokines in the Nervous System, edited by Nancy J. Rothwell. © 1996 R.G. Landes Company.

we describe the modulation of IL-1 receptors following endotoxin treatment.

MOLECULAR BIOLOGICAL CHARACTERISTICS OF IL-1 RECEPTORS

CLONING OF IL-1 RECEPTORS AND RELATED GENES

IL-1 exerts its biological actions through specific receptors. Two major classes of IL-1 receptors have been cloned from both mouse and human and designated as type I (IL-1RI)[29,30] and type II (IL-1RII)[31] IL-1 receptors. IL-1RI and IL-1RII share ~30% homology (Table 3.1). Both IL-1RI and IL-1RII are glycoproteins with molecular weights of IL-1α~80 and ~68 kDa, respectively. These two receptors bind IL-1α, IL-1β and IL-1ra but with different potencies.[30,31] While IL-1RI binds all three ligands with comparable affinity, IL-1RII has much higher affinity for IL-1β than IL-1α or IL-1ra. Additional members of the IL-1 receptor family have been cloned in the last few years. These include mouse ST2L,[32] its rat counterpart Fit-1,[33] an IL-1 receptor related protein (IL-1R-rp)[34] and IL-1 receptor accessory protein (IL-1RAcP).[35] While the functions of ST2, Fit-1 and IL-1R-rp remain unclear, IL-1RAcP was proposed to associate with IL-1RI and facilitate high affinity binding of IL-1 to IL-1RI.[35] Rat homologues of IL-1RI,[36] IL-1RII,[37] IL-1R-rp[34] and IL-1RAcP[38] have also been cloned. All members of the IL-1 receptor gene family have a similar gene structure. Both IL-1RI and IL-1RII and other members of the IL-1 receptor gene family have an immunoglobulin-like extracellular domain. While IL-1RI has a long intracellular domain (213 amino acids),[30] IL-1RII possesses a very short intracellular domain (29 amino acids).[31] In addition to the three putative disulfide bonds which are common to both IL-1RI and IL-1RII, IL-1RII possesses a fourth putative disulfide bond.

In addition to the transmembrane IL-1 receptors mentioned above, a vaccinia virus gene B15 was demonstrated to encode a soluble IL-1 receptor.[39] Soluble receptor mRNAs for rat IL-1RI,[40] human and monkey IL-1RII,[41] ST2,[42] Fit-1,[33] rat IL-1RAcP (C. Liu et al, unpublished), and mouse IL-1RAcP[35] have also been cloned. These soluble receptors for the most part demonstrate comparable binding characteristics to their transmembrane counterparts. We have recently cloned a cDNA from human Raji cells and COS1 cell for an alternatively processed IL-1RII mRNA.[41] Expression of this cDNA in COS7 cells demonstrated that it encodes a soluble IL-1 receptor with ligand-binding properties comparable to IL-1RII.[41]

SIGNAL TRANSDUCTION MEDIATED BY IL-1 RECEPTORS

Several lines of evidence indicate that IL-1RI is the primary receptor that transduces IL-1's signaling.[43,44] Through IL-1R, IL-1 activates c-Jun, c-Fos,[45-47] NF-κB[43,48] and many other signal transduction events.

Table 3.1. Percent homology between members of IL-1 receptor gene family

		IL-1RI			IL-1RII				IL-1R-rp		IL-1RAcP		ST2/Fit-1	
		Human	Mouse	Rat	Human	Monkey	Mouse	Rat	Human	Rat	Mouse	Rat	Mouse	Rat
IL-1RI	Human	100												
	Mouse	68.1	100											
	Rat	67.0	85.6	100										
IL-1RII	Human	27.1	29.5	29.8	100									
	Monkey	27.5	31.0	31.3	93.0	100								
	Mouse	25.5	26.4	29.5	60.0	58.7	100							
	Rat	25.9	26.4	29.5	60.8	59.0	90.0	100						
IL-1R-rp	Human	39.6	38.0	38.7	26.2	27.3	27.0	26.5	100					
	Rat	36.6	34.8	35.7	26.9	27.2	25.8	26.5	66.8	100				
IL-1RAcP	Mouse	23.7	25.6	26.1	22.4	23.1	22.2	23.0	23.6	24.4	100			
	Rat	23.4	25.6	25.3	22.5	22.9	22.5	23.3	23.7	24.8	95.1	100		
ST2/Fit-1	Mouse	27.2	27.6	26.9	22.5	22.4	24.1	24.1	27.4	26.7	25.3	25.4	100	
	Rat	27.2	28.1	26.6	21.9	21.6	22.5	22.2	27.7	27.7	24.5	24.5	85.7	100

Protein amino acid sequence between different members of IL-1 receptor gene family were compared using DNASTAR protein sequence analysis program: Lipman-Pearson method.

The immediate early events of IL-1 receptor mediated signal transduction are not entirely clear. The activated IL-1 receptor has also been proposed to initiate the sphingomyelin signal transduction pathway.[49] Very recently, an IL-1 receptor associated kinase (IRAK) was identified and cloned.[50] IL-1 was reported to promote the association of IRAK with IL-1RI and phosphorylation of IRAK in 293 cells that overexpress IL-1RI.[50] The amino acid sequence of IRAK shares homology to that of Pelle, a protein kinase that is essential for the activation of a NF-κB homolog in *Drosophila*.[51,52]

IL-1RII has been postulated to function primarily as a decoy receptor for IL-1,[44] but the function of IL-1RII in IL-1 signaling cannot be completely excluded. A monoclonal antibody to IL-1RII blocks the thermogenic and pyrogenic effects of IL-1β in the brain but not IL-1α[53] suggesting the possible presence of IL-1RII mediated IL-1 signaling in the brain. The signal transduction of IL-1β through IL-1RII may be accomplished by IL-1RII complexing with another transmembrane protein as was recently demonstrated for IL-1RI.[35]

RADIOLIGAND BINDING CHARACTERISTICS OF IL-1 RECEPTORS IN BRAIN

REGIONAL DISTRIBUTION OF [^{125}I]IL-1 BINDING SITES IN CNS: HOMOGENATE STUDIES

The regional distribution of binding sites for [^{125}I]IL-1α was examined in homogenates of discrete areas of mouse CNS in order to identify a brain area(s) that may be ideally suited for subsequent characterization of the receptor. The highest density of binding sites in mouse CNS was present in the hippocampus.[54] Progressively lower, but significant densities of binding sites were detected in cerebral cortex, cerebellum, olfactory bulb, striatum, spinal cord, hypothalamus and medulla oblongata. Therefore, mouse hippocampus was used for the remainder of the characterization studies.

SPECIES DIFFERENCES IN IL-1 RECEPTORS

The nucleotide and amino acid sequence identity of IL-1RI between mouse human and rat is very high (i.e., >65%) (Table 3.1). Comparable high sequence identity of IL-1RII between species is also present (Table 3.1). Binding of radioiodinated forms of various IL-1s were compared in mouse, rat, guinea pig and rabbit hippocampus.[54,55] In contrast to moderate to high levels of binding in mouse and rabbit tissues, [^{125}I]IL-1α and [^{125}I]IL-1ra binding to rat or guinea pig tissues was barely within the range of sensitivity of the assay. In addition, various radioligands including human, mouse or rat forms of IL-1 and IL-1ra were used to label IL-1 binding sites in mouse and rat tissues. [^{125}I]hIL-1α, [^{125}I]hIL-1ra, [^{125}I]rIL-1β and [^{125}I]rIL-1ra showed high specific binding with progressively lower binding evident with [^{125}I]mIL-1β and [^{125}I]hIL-1β in mouse tissues. No specific IL-1 binding was found in rat hippocampus using any of the radioligands described above. The absence of specific [125I]IL-1α and [125I]IL-1ra binding in rat hippocampus does not necessarily suggest a lack of IL-1 receptors, since recombinant human IL-1 in rats alters sleep,[56] induces anorexia,[57] and induces adrenocorticotropic hormone (ACTH) release.[58,59] In addition, IL-1ra reduced the severity of experimental enterocolitis and lipopolysaccharide-induced pulmonary inflammation in rat.[60] Recent data suggest a multiplicity of IL-1 receptors (see above), and the possibility exists that the radioligands used in the present study (recombinant human [125I]IL-1α and [125I]IL-1ra) only label a subtype of these receptors which is present in some species including mouse, rabbit, human and monkey (unpublished data) but is absent in species such as rat and guinea pig. A very recent report demonstrated that high densities of specific rat [125I]IL-1β binding sites were localized in the anterior as well as the posterior pituitary and in the choroid plexus in the rat tissues by quantitative autoradiography.[61] Additional studies may be necessary in resolving these species differences.

KINETIC AND PHARMACOLOGICAL CHARACTERISTICS OF BRAIN IL-1 RECEPTORS

The concentration-dependent binding of [125I]IL-1α and [125I]IL-1ra to mouse hippocampus under equilibrium conditions was examined.[54,62] Specific [125I]IL-1α or [125I]IL-1ra binding was saturable and of high affinity. Scatchard analysis of the saturation data showed comparable high affinity binding (K_D: 60-120 pM for [125I]IL-1α and 20-30 pM for [125I]IL-1ra) in mouse brain.

The pharmacological characteristics of the [125I]IL-1α and [125I]IL-1ra binding sites were examined by determining the relative potencies of IL-1-related and -unrelated peptides in displacing specifically bound [125I]IL-1α and [125I]IL-1ra in homogenates of mouse brain.[54,62] The results of the homogenate studies with [125I]IL-1α and [125I]IL-1ra are summarized in Table 3.2. IL-1α and IL-1ra were more potent than IL-1β which, in turn, was more potent than its weak analogs IL-1β+ and IL-1βc.[18] Corticotropin releasing factor (CRF) and tumor necrosis factor-α (TNF-α) (at concentrations up to 100 nM) had no effect on [125I]IL-1α or [125I]IL-1ra binding. The relative inhibitory potencies of IL-1α, IL-1β, IL-1β+ and IL-1βc[18], for the most part, paralleled their bioactivities in a murine thymocyte co-stimulation assay.[63]

RELATIVE DISTRIBUTION OF IL-1 RECEPTOR AND IL-1 RECEPTOR mRNA IN MOUSE AND RAT BRAIN

DISTRIBUTION OF [125I]IL-1 BINDING SITES IN MOUSE BRAIN: RECEPTOR AUTORADIOGRAPHIC STUDIES

[125I]IL-1α or [125I]IL-1ra were utilized as radioligands for autoradiographic studies. Overall, very low densities of [125I]IL-1α and [125I]IL-1ra binding sites were present throughout the brain. Very high

Table 3.2. Pharmacological specificity of [^{125}I]IL-1α-and [^{125}I]IL-1ra binding to mouse hippocampus

Peptide	K$_i$ (pM)		Biological Activity (units/mg)
	[^{125}I]IL-1α	[^{125}I]IL-1ra	
IL-1α	55 ± 18	70 ± 10	3.0 x 10^7
IL-1ra	N.D.	119 ± 63	No activity
IL-1β	76 ± 20	1798 ± 234	2.0 x 10^{7*}
IL-1β$^+$	2940 ± 742	N.D.	1.0 x 10^6
IL-1β^{c18}	N.D.	2008 ± 350	N.D.
TNF	>100,000	>100,000	8.0 x 10^2
CRF	>100,000	>100,000	0.0

Peptides at 3-10 concentrations were incubated with approximately 100 pM [^{125}I]IL-1α and 40 pM [^{125}I]IL-1ra for 120 min at room temperature. All assays were conducted in triplicate in three separate experiments. K$_i$ (inhibitory binding-affinity constant) values were obtained from competition curve data analyzed using the computer program LIGAND.[64] Biological activity data were obtained in a murine thymocyte assay.[63] * Of note, IL-1β used in the [^{125}I]IL-1ra experiments had lower biological activity than IL-1β used in the [^{125}I]IL-1α experiments. Abbreviations: CRF, rat/human corticotropin-releasing factor; TNF, human recombinant tumor necrosis factor-α N.D., not determined.

densities and a discrete localization of IL-1 receptors were evident in the hippocampal formation and in the choroid plexus[54,62] (Fig. 3.1). Within the hippocampus, IL-1 receptors were present in the molecular and granular layers of the dentate gyrus and were virtually absent in the CA1 to CA3 pyramidal region. The pharmacological binding characteristics of [^{125}I]IL-1α were similar in the dentate gyrus and in choroid plexus and, for the most part, comparable to the characteristics seen in homogenates of hippocampus. IL-1α and IL-1β inhibited binding in both areas with comparable potencies (IC$_{50}$ values of less than 100 pM). The weak IL-1β analog, IL-1β$^+$ inhibited 80-85% of the specific binding at the high concentration of 50 nM, while comparable concentrations of TNF-α and CRF were essentially ineffective in inhibiting [^{125}I]IL-1α binding. There was an absence of specific [^{125}I]IL-1ra or [^{125}I]IL-1α binding in the hypothalamus, cerebral cortex and other brain areas.

Excitatory amino acid lesions of the hippocampus were utilized to determine if the [^{125}I]IL-1α binding sites were localized to specific neuronal systems. In the hippocampus, intrinsic neurons were destroyed by local injection of quinolinic acid. This treatment abolished [^{125}I]IL-1α binding sites in both the granular and molecular layers of the dentate gyrus, indicating that IL-1 receptors were localized to intrinsic neurons.[64]

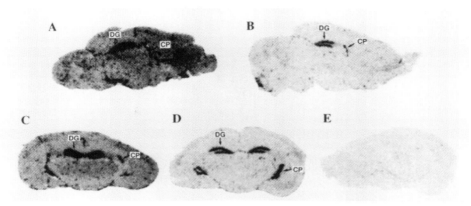

Fig. 3.1. Autoradiographic localization of [^{125}I]IL-1α and [^{125}I]IL-1ra binding in mouse brain cut in sagittal (top) and coronal (bottom) planes. The tissues were incubated for 120 min with 40 pM of [^{125}I]IL-1ra (B, D and E) or 100 pM of [^{125}I]IL-1α (A and C). The images were computer generated using autoradiograms on Hyperfilm. The darker areas in autoradiograms (A-D) correspond to brain regions displaying higher densities of binding. In E, note the absence of specific [^{125}I]IL-1ra binding in a section adjacent to D (i.e., blank) coincubated with 100 nM IL-1α. Abbreviations: DG, dentate gyrus; CP, choroid plexus. (Reproduced with permission from Takao et al[62]).

RELATIVE DISTRIBUTION OF MRNA FOR IL-1RI AND IL-1RAcP: RNASE PROTECTION STUDIES IN RAT BRAIN

Basal levels of IL-1RI mRNA are normally not detectable by Northern blot but detectable in microdissected rat brain regions by RNase protection assays.[38,65] Overall, there was a low level of IL-1RI mRNA expression in rat brain. A 378 bp protected fragment corresponding to the membrane-bound form of IL-1RI was detected in a variety of brain areas including thalamus, neostriatum, hippocampus, medulla, midbrain and cerebral cortex.[65] Expression of mRNA encoding the soluble form of the IL-1RI was not detected in rat brain.[65] In contrast to the low mRNA expression levels of IL-1RI, the mRNA of IL-1RAcP, which has been proposed to associate with IL-1RI and facilitate IL-1RI to bind IL-1, was expressed at much higher levels in the brain.[35,38] IL-1RAcP mRNA is very highly expressed in hippocampus, hypothalamus, cerebral cortex and cerebellum (Fig. 3.2).

RELATIVE DISTRIBUTION OF MRNA FOR IL-1RI IN MOUSE BRAIN: *IN SITU* HYBRIDIZATION STUDIES

In situ hybridization histochemistry was used to investigate the distribution of cells expressing IL-1RI receptor mRNA in mouse brain. The strongest autoradiographic signal in the forebrain was found in the hippocampal formation where an intense autoradiographic signal was observed over the granule cell layer of the dentate gyrus, and a

Fig. 3.2. Tissue distribution of rat IL-1RAcP mRNA. Data presented are representative of three separate experiments with similar results. Five micrograms of total RNA from different tissues was used for RNase protection assays for rat IL-1RAcP mRNA (A). One microgram of total RNA from different samples was assayed for rat β-actin mRNA in parallel as control (B). The protected fragments were then run on to a 5% denaturing acrylamide gel and exposed to a X-ray film overnight. The relative IL-1RAcP mRNA levels between samples were expressed as a ratio to β-actin mRNA signal which was quantitated by phosphoimager (Bio-Rad) (C). Hy: hypothalamus; Hi: hippocampus; Ctx: cerebral cortex; Cb: cerebellum; Liv: liver; Lun: lung; Thy: thymus; Sp: spleen. (Reproduced with permission from Liu et al[38]).

A: IL-1RAcP

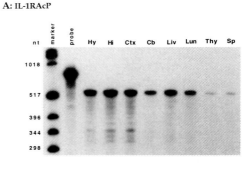

B: Actin

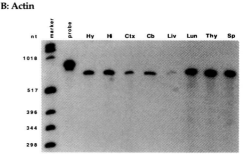

C: IL-1RAcP/Actin

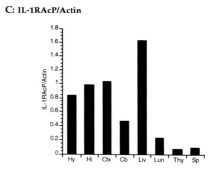

weak to moderate autoradiographic signal was observed over the pyramidal cell layer of the hilus and the CA3 region.[66,67] The autoradiographic signal in the hypothalamic paraventricular nucleus and most aspects of median eminence was comparable to background.[67] An intense autoradiographic signal was observed over all aspects of the midline raphe system.[67] A moderate to dense autoradiographic signal was observed over sensory neurons of the mesencephalic trigeminal nucleus.[67] A weak to moderate autoradiographic signal was found over the Purkinje cell layer of the entire cerebellar cortex. A dense autoradiographic signal was observed over the choroid plexus in the lateral, third, and fourth ventricles.[67] An intense autoradiographic signal was found over endothelial cells of postcapillary venules throughout the CNS, both in the parenchyma and at the pial surface.[67]

The hippocampus has been suggested to be involved in thermoregulation, since it has been shown that hippocampal stimulation may directly influence thermoresponsive neurons in the preoptic area.[68,69] Moreover, IL-1 might exert at least some of its central effects on the hypothalamic-pituitary axes at the level of the hippocampus, as has been postulated for glucocorticoids and regulation of HPA axis.[70,71] This possibility seems particularly appealing given the relative absence of type I IL-1 receptor mRNA in ether the paraventricular nucleus of the hypothalamus or median eminence although IL-1 has been shown to stimulate release of CRF from perfused rat hypothalami,[72,73] and to produce an increase in plasma ACTH following microinjection into the median eminence in vivo.[74]

RELATIVE DISTRIBUTION OF mRNA FOR IL-1RI AND IL-1RAcP IN RAT BRAIN: IN SITU HYBRIDIZATION STUDIES

In situ hybridization studies indicated dramatic species differences between the localization of IL-1RI in rat and mouse brain. Overall, rat IL-RI mRNA was predominantly expressed over barrier-related cells, including the leptomeninges, non-tanycytic portions of epedema, the choroid plexus, and vascular endothelium (Fig. 3.3).[38,65] In contrast to the intense neuronal localization seen in mouse brain, only very low to moderate levels of the IL-1RI mRNA were detected in neuronal groups such as the basolateral nucleus of the amygdala, the arcuate nucleus of the hypothalamus, the trigeminal and hypoglossal motor nuclei, and the area postrema.[65] In another in situ hybridization study, it was reported that IL-1RI mRNA was expressed more extensively in other rat brain regions such as the anterior olfactory nucleus, medial and posterior thalamic nuclei, ventromedial hypothalamic nucleus, median eminence, and Purkinje cells of the cerebellum and hippocampus.[75] The reasons for the dramatic species differences in view of the close sequence identity between mouse and rat IL-1RI are not entirely clear.

Compared with the regional expression of IL-1RI mRNA in rat brain, rat IL-1RAcP mRNA is expressed at high levels in rat brain and most highly expressed in hippocampus. Within the hippocampus, IL-1RAcP mRNA is expressed primarily in the dentate gyrus of the hippocampus but not in the adjacent CA fields (Fig. 3.3). This expression is predominantly confined to dentate granule cells within the dorsal hippocampus and also includes some interneurons in the ventral dentate gyrus. In situ hybridization analysis indicated a low to negligible level of neuronal expression for IL-1RAcP mRNA in other brain regions, including the hypothalamus. Furthermore, like IL-1RI, IL-1RAcP mRNA was evident in non-neuronal cells, such as blood

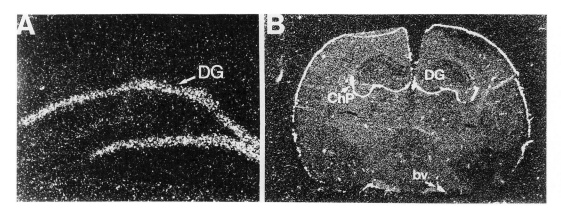

Fig. 3.3. Comparative localization of IL-1 RAcP mRNA and IL-1R1 mRNA expression in rat brain. A, High power magnification showing the high expression level of IL-1 RAcP mRNA in the dentate gyrus of the hippocampus (DG). B, Low power magnification of a coronal section of brain showing the localization of IL-1R1 mRNA predominantly in non-neuronal cells including blood vessels (bv) and the choroid plexus (ChP). In B, note the absence of IL-1R1 mRNA localization in the dentate gyrus. (Reproduced with permission from Liu et al[38]).

vessels and parenchymal cells which account for higher levels of mRNA expression in RNase protection studies in brain areas such as the cerebral cortex, hypothalamus and cerebellum.

The marked discrepancy between the localization of IL-1RAcP and IL-1R1 mRNA in rat brain suggests that the association between the two proteins may not be present in rat brain. Radioligand binding studies in rat brain further support the absence of a complex between the two proteins that may be required for high affinity binding. There is a virtual absence of high affinity binding of various forms of [125I]-IL1 in rat brain comparable to that seen for prototypic type I IL-1 receptors in EL-4 6.1 cells.[55] Only low affinity [125I]human IL-1β binding in rat hypothalamic and cerebral cortical homogenates[76] and [125I]mouse IL-1α binding in slide-mounted rat brain sections[77] have been reported. This apparent lack of association of the two proteins in brain may be more evident in rats than in some other species. High affinity [125I]-IL1 binding with kinetic and pharmacological characteristics of type I receptors is seen in mouse hippocampus[54,62] where, in contrast to the rat, we have observed a co-expression of murine type I IL-1 and IL-1RAcP mRNA in the dentate gyrus of the hippocampus[67] (D. Chalmers et al, unpublished).

The presence of the IL-1RAcP in brain areas which show an absence of type I IL-1 receptors suggest additional functions for the newly identified protein in the rat. IL-1RAcP may be associated with other novel IL-1 receptors or may function as a separate receptor entity. The fact that IL-1RAcP has very similar extracellular and intracellular signaling domain characteristics to other known members of the IL-1 receptor family including the type I IL-1 receptor and the Fit-1/ST2[32,33] may define it as a separate receptor. In addition, IL-1RAcP may interact in a manor analogous to type I IL-1 receptors with novel receptors or existing proteins such as Fit-1 that show only low affinity binding of IL-1β.[78] Alternatively, IL-1RAcP may be a primary receptor for ligands in addition to IL-1.

MODULATION OF IL-1 RECEPTORS IN BRAIN AND PERIPHERY FOLLOWING ENDOTOXIN TREATMENT

In an attempt to define the regulation of IL-1 and IL-1 receptors in the mouse brain-endocrine-immune axis, we measured tissue levels of IL-1β using an ELISA and iodine-125-labeled recombinant human interleukin-1α ([^{125}I]IL-1α) binding in hippocampus, hypothalamus, pituitary, epididymis, testis, liver and spleen after ip injection of the bacterial endotoxin, lipopolysaccharide (LPS).[79] Basal IL-1β levels were detectable in all the tissues examined. IL-1β levels were dramatically increased in the peripheral tissues (pituitary, testis and spleen) at 2-6 h after a single LPS injection, however, no significant changes were observed in brain (hippocampus and hypothalamus).[79] [^{125}I]IL-1α binding in the pituitary gland, liver, spleen and testis was significantly decreased at 2 h following a single administration of both low (30 μg LPS/mouse) and high (300 μg LPS/mouse) doses of endotoxin.[80] On the other hand, [^{125}I]IL-1α binding in the hippocampus was not significantly altered at 2 h by low dose of LPS and was only significantly decreased by high dose administration of LPS (300 μg/mouse).[80] The acute (2 h) decrease in [^{125}I]IL-1α binding in peripheral tissues most likely represents primarily an occupancy rather than a down-regulation of the IL-1 receptor by high concentrations of tissue IL-1 since the decrease in binding was transient rather than long-lasting.[79] Furthermore, the rapid rate of receptor turnover required to produce the 2 h decreases in [^{125}I]IL-1α binding is highly unlikely. In contrast to the observations in the periphery, IL-1β levels in the hippocampus or hypothalamus and [^{125}I]IL-1α binding remained unchanged in the hippocampus following acute low-dose LPS administration. The differences in LPS-induced regulation of [^{125}I]IL-1α binding in the hippocampus and peripheral tissues may be due to the unaltered levels of IL-1 in the hippocampus when compared to peripheral tissues[79] and/or the lack of access to the hippocampus of circulating IL-1 due to the presence of the blood-brain barrier.

In order to evaluate if activation of IL-1 in brain may require more sustained exposure to endotoxin, we examined the effects of two injections of LPS at 0 and 12 h. Following two LPS injections (at 0 and 12 h), dramatic increases in IL-1β concentrations in the hypothalamus, hippocampus, spleen and testis were observed at 2 h after the second LPS injection; a small but statistically nonsignificant change was evident in the pituitary. The concentrations of IL-1β in hypothalamus

and hippocampus returned to basal levels by 6 h, while the levels remained elevated in the testis and spleen (but significantly lower than
those seen at 2 h) at 6 h and returned to control levels by 24 h after
the second injection.[79] The largest relative LPS-induced increases in
IL-1β were observed in the testis (27-fold), with lower increases seen
in the hippocampus (6-fold), spleen (4.5-fold) and hypothalamus
(2.2-fold).

In order to determine whether regulation of [^{125}I]IL-1α binding in
the brain is simply dependent on the total dose of LPS administered
or whether more sustained exposure to endotoxin may be required, we
measured [^{125}I]IL-1α binding in the hippocampus and testis at 2 h
following a single injection (60 μg/mouse) or following two injections
(30 μg/mouse each at 0 and 12 h) of endotoxin.[80] Following one injection of LPS (60 μg/mouse), [^{125}I]IL-1α binding was significantly
decreased by ~50% in the testis. While [^{125}I]IL-1α binding in hippocampus had tendency to decline after one injection of LPS (60 μg/mouse),
no statistically significant differences between saline- and LPS-induced
animals were observed (Fig. 3.4). In contrast, dramatic decreases in
[^{125}I]IL-1α binding were seen after two injections of 30 μg LPS/mouse
in both the hippocampus and testis[80] (Fig. 3.4). These data suggest
that activation of IL-1 in the brain may require more sustained exposure to endotoxin than in peripheral tissues. The sustained exposure
to endotoxin may induce not only more IL-1 but also other cytokines
such as TNF-α and interleukin-6,[74,81-83] speculating that these cytokines affect IL-1 production in the brain which reduces IL-1 receptors
in the hippocampus. A previous study has demonstrated significant
reductions in [^{125}I]IL-1α binding in the dentate gyrus of the hippocampus, in the choroid plexus, but not in the anterior pituitary at 20 h
after a single injection of LPS.[84] The decreases in [^{125}I]IL-1α binding
in the various tissues were evident at a time (12 h after the second
LPS injection) when the levels of IL-1 had almost returned to basal
concentrations,[79] suggesting that the decreases in [^{125}I]IL-1α binding
were most likely due to a decrease in the density of the receptors rather
than simply due to an occupancy of the receptor by the endogenous
ligand. Further evidence for this contention was provided by autoradiographic studies which demonstrated significant reductions in the density
of [^{125}I]IL-1α binding in both brain and peripheral tissues in LPS-
treated mice when compared to saline-injected animals.[79]

The modulation of rat IL-1RAcP and rat IL-1RI mRNA in hippocampus and liver was examined following a two-injection treatment
regimen of LPS. The results clearly indicated that while LPS treatment down regulated both IL-1RI mRNA and IL-1RAcP mRNA in
liver, no significant alterations were seen in hippocampus following
this treatment regimen[38] (Fig. 3.5). The lack of modulation of IL-1RAcP
in brain following LPS treatment was confirmed in in situ hybridization studies. Treatment of mice with IL-1α for 4 h resulted in a re-

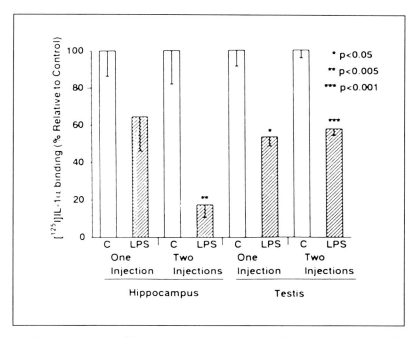

Fig. 3.4. Comparison of the one and two injection(s) of lipopolysaccharide (LPS) on [¹²⁵I]IL-1α binding in the hippocampus and testis. Mice were injected with LPS (60 μg/mouse, n = 5) at time 0 and sacrificed 2 h after the LPS injection or injected with LPS (30 μg/mouse, n = 6) at time 0 and 12 h and sacrificed 12 h after the second LPS injection. Data, expressed as a percent of specific binding in saline-injected controls, represent mean ± SEM (n = 5 in one LPS injection and n = 10 in two LPS injections). *, ** and *** represent significant differences at p < 0.05, p < 0.005, p < 0.001, respectively from saline-injected controls as determined by one-way analysis of variance followed by Fisher's protected least squares difference test. (Reproduced with permission from Takao et al[80]).

duction in murine IL-1RAcP mRNA in the liver[35] suggesting that the effects of LPS treatment to downregulate liver IL-1RAcP and IL-1R1 mRNA in rats is most likely mediated by the endogenous cytokine.

In contrast to the periphery, endotoxin treatment did not modulate either IL-1RAcP and IL-1R1 mRNA in rat hippocampus[38] (Fig. 3.5). The LPS treatment used in the present study may have been insufficient to modulate IL-1 levels in rat brain. Alternatively, the IL-1RAcP and IL-1R1 mRNA in the liver and brain may be differentially modulated by IL-1. Treatment of mice with IL-1α has been reported to produce differential effects on mRNA levels of IL-1RAcP in different tissues with decreases noted in liver, no change in brain and increases seen in the lung and spleen.[35] It is unclear at the present time if these differential effects of endotoxin treatment may relate to the membrane and soluble forms of the of IL-1RAcP that have been reported to exist

A

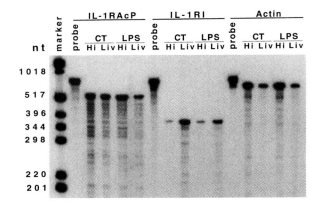

B

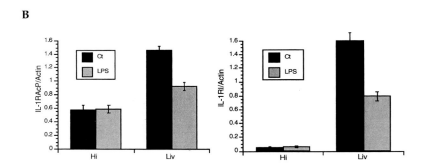

Fig. 3.5. IL-1RAcP and IL-1RI mRNA regulation by endotoxin treatment. A. 5 µg of total RNA from different regions of LPS treated or control rats was assayed for IL-1RAcP and IL-1RI mRNA expression levels by RNase protection assay. One microgram of total RNA from each sample was assayed in parallel for β-actin mRNA. B. The experiment was done in triplicate and the relative IL-1RAcP and IL-1RI mRNA expression levels were expressed as ratios to β-actin. Abbreviations: CT: vehicle-treated group; Hi: hippocampus; Liv: liver; LPS: lipopolysaccharide-treated group. *, significantly different from vehicle-treated controls at p < 0.05. (Reproduced with permission from Liu et al[38]).

in varying proportions in brain and peripheral tissues[35] or in species differences that may exist.

SUMMARY

Two major types of functional IL-1 receptors have been identified, cloned and characterized. In addition to the two IL-1 receptors (IL-1RI and IL-1RII), several other cDNAs for the IL-1 receptor family, including IL-1R-rp, IL-1AcP, ST2 and Fit-1, have also cloned. Those members of IL-1 receptor family all share about 20% or greater homology to each other. IL-1AcP has been proposed to associate with IL-1RI and facilitate its high-affinity binding to IL-1. Soluble receptor cDNAs for the members of IL-1 receptor gene family have also been cloned. Although different lines of evidence suggest that IL-1RI is the "sole" IL-1 receptor that transduces IL-1's signaling and IL-1RII only acts as a decoy receptor, some other results indicate that IL-1RII may still play a important role in IL-1 signaling.

IL-1 receptors were identified, characterized and localized in mouse brain and using [^{125}I]IL-1α and [^{125}I]IL-1ra as radioligands. [^{125}I]IL-1 binding in mouse brain was linear with membrane protein concentration, saturable, reversible and of high affinity. The binding sites for [^{125}I]IL-1 exhibited a pharmacological specificity for IL-1 and its analogs in keeping with the relative biological potencies of the compounds in the thymocyte proliferation assay. Receptor autoradiography studies demonstrated high densities and a discrete localization of IL-1 receptors and receptor mRNA, respectively, in the dentate gyrus of the hippocampus and in choroid plexus. IL-1β concentrations and IL-1 receptors were reciprocally modulated in brain and in the periphery following endotoxin treatment. The degree of modulation was dependent on the dose, frequency of administration as well as the species under investigation (i.e., rat vs. mouse). The demonstration of the presence of high-affinity receptors that are discretely localized in brain provides further support for the proposed role for IL-1 in modulating CNS function.

REFERENCES

1. Dinarello CA. Interleukin-1 and interleukin-1 antagonism. Blood 1991; 77:1627-52.
2. Mizel SB. The interleukins. FASEB J 1989; 3:2379-88.
3. Oppenheim JJ, Kovacs EJ, Matsushima K et al. There is more than one interleukin 1. Immmunol Today 1986; 7:45-6.
4. Fontana A, Kristensen F, Dubs R et al. Production of prostaglandin and interleukin-1 like factor by cultured astrocytes and glioma cells. J Immunol 1982; 129:2413-9.
5. Giulian D, Baker TJ, Shih LC et al. Interleukin 1 of the central nervous system is produced by ameboid microglia. J Exp Med 1986; 164:594-604.
6. Giulian D, Young DG, Woodward J et al. Interleukin-1 is an astroglial growth factor in the developing brain. J Neurosci 1988; 8:709-14.
7. Giulian D, Lachman LB. Interleukin-1 stimulates astroglial proliferation after brain injury. Science 1985; 228:497-9.
8. Perry VH, Brown MC, Gordon S. The macrophage response to central and peripheral nerve injury: A possible role for macrophage in regeneration. J Exp Med 1987; 165:1218-23.
9. Fontana A, Weber E, Dayer JM. Synthesis of interleukin-1/endogenous

pyrogen in the brain of endotoxin treated mice: a step in fever induction. J Immunol 1984; 133:1696-8.

10. Lue FA, Bail M, Gorczynski R et al. Sleep and interleukin-1-like activity in cat cerebrospinal fluid. Int J Neurosci 1988; 42:179-83.

11. Mustafa MM, Lebel MH, Ramio O et al. Correlation of interleukin-1 β and cachectin concentrations in cerebrospinal fluid and outcome from bacterial meningitis. J Pediatr 1989; 115:208-13.

12. Farrar WL, Hill JM, Harel-Bellan A et al. The immune logical brain. Immunol Rev 1987; 100:361-8.

13. Heier E, Ayala J, Denefle A et al. Brain macrophages synthesize interleukin-1 and interleukin-1 mRNA in vitro. J Neurosci Res 1988; 21:391-7.

14. Breder CD, Dinarello CA, Saper CB. Interleukin-1 immunoreactive innervation of the human hypothalamus. Science 1988;240:321-4.

15. Breder CD, Saper CB. Interleukin-1β-like immunoreactive innervation in the human central nervous system. Soc Neurosci Abstr 1989; 15:715.

16. Krueger JM, Walter J, Dinarello CA et al. Sleep-promoting effects of endogenous pyrogen (interleukin-1). Am J Physiol 1984;246:R994-9.

17. Tobler I, Borbely AA, Schwyzer M et al. Interleukin-1 derived from astrocytes enhances slow wave activity in sleep EEG of the rat. Eur J Pharmacol 1984; 104:191-2.

18. McCarthy DO, Kluger MJ, Vander AJ. Suppression of food intake during infection: is interlexukin-1 involved? Am J Clin Nutr 1985; 42:181-4.

19. Nakamura H, Nakanishi S, Kita A et al. Interleukin-1 induces analgesia in mice by a central action. Eur J Pharmacol 1988; 159:49-54.

20. Kampschmidt RF. The numerous postulated biological manifestations of interleukin-1. J Leukoc Biol 1984; 36:341-55.

21. Blatteis CM. Neural mechanisms in the pyrogenic and acute-phase responses to interleukin-1. Int J Neurosci 1988; 38:223-32.

22. Dascombe MJ, Hardwick A, Lefeuvre RA et al. Impaired effects of interleukin-1 beta on fever and thermogenesis in genetically obese rats. Int J Obes 1989; 13:367-73.

23. Busbridge NJ, Dascombe MJ, Tilders FJ et al. Central activation of thermogenesis and fever by interleukin-1 beta and interleukin-1 alpha involves different mechanisms. Biochem Biophys Res Commun 1989; 162:591-6.

24. Sundar SK, Cicerpial MA, Kilts C et al. Brain IL-1-induced immunosuppression occurs through activation of both pituitary-adrenal axis and sympathetic nervous system by corticotropin-releasing factor. J Neurosci 1990; 10:3701-6.

25. Berkenbosch F, Van Oers J, Del Rey A et al. Corticotropin-releasing factor-producing neurons in the rat activated by interleukin-1. Science 1987; 238:524-6.

26. Sapolsky R, Rivier C, Yamamoto G et al. Interleukin-1 stimulates the secretion of hypothalamic corticotropin-releasing factor. Science 1987; 238:522-4.

27. Uehara A, Gottschall PE, Dahl RR et al. Interleukin-1 stimulates ACTH

release by an indirect action which requires endogenous corticotropin-releasing factor. Endocrinology 1987; 121:1580-2.

28. Rivier C, Vale W. In the rat, interleukin-1α acts at the level of the brain and the gonads to interfere with gonadotropin and sex steroid secretion. Endocrinology 1989; 124:2105-9.

29. Sims JE, March CJ, Cosman D et al. cDNA expression cloning of the IL-1 receptor, a member of the immunoglobulin superfamily. Science 1988; 241:585-9.

30. Sims JE, Acres RB, Grubin C. Cloning the interleukin 1 receptor from human T cells. Proc Natl Acad Sci USA 1989; 86:8946-50.

31. McMahan CJ, Slack JL, Mosley B et al. A novel IL-1 receptor, cloned from B cells by mammalian expression, is expressed in many cell types. EMBO 1991; 10:2821-32.

32. Yanagisawa K, Takagi T, Tsukamoto T et al. Presence of a novel primary response gene ST2L, encoding a product highly similar to the interleukin 1 receptor type 1. FEBS Lett 1993; 318:83-7.

33. Bergers G, Reikerstorfer A, Braselmann S et al. Alternative promoter usage of Fos-resposive gene Fit-1 generates mRNA isoforms coding for either secreted or membrane-bound proteins related to the IL-1 receptor. EMBO 1994; 13:1176-88.

34. Lovenberg TW, Crowe PD, Liu C et al. Cloning of a cDNA encoding a novel interleukin-1 receptor related protein (IL-1R-rp). J Neuroimmunol 1996; (in press).

35. Greenfeder SA, Nunes P, Kwee L et al. Molecular cloning and characterization of a second subunit of the interleukin 1 receptor complex. J Biol Chem 1995; 270: 13757-65.

36. Hart RP, Liu C, Shadiack AM et al. An mRNA homology to interleukin-1 receptor type I is expressed in cultured rat sympathetic ganglia. J Neuroimmunol 1993; 44:49-56.

37. Bristulf J, Gatti S, Malinowsky D et al. Interleukin-1 stimulates the expression of type I and type II interleukin-1 receptors in the rat insulinoma cell line Rinm5F; sequencing a rat type II interleukin-1 receptor cDNA. Eur Cytokine Netw 1994; 5:319-330.

38. Liu C, Chalmers DT, Maki R et al. Rat homolog of mouse interleukin-1 receptor accessory protein: cloning, localization and modulation studies. J Neuroimmunol 1996; 66:41-8.

39. Alcami A, Smith GL. A soluble receptor for interleukin-1 beta encoded by vaccinia virus: a novel mechanism of virus modulation of the host response to infection. Cell 1992; 71:153-67.

40. Liu C, Ganea D, Hart RP. A soluble IL-1 receptor is expressed in cultured rat sympathetic ganglia from the type I IL-1 receptor gene. Soc Neurosci Abstr 1993; 19:95.

41. Liu C, Hart RP, Liu X-J et al. Cloning and characterization of an alternatively processed human type II interleukin-1 receptor mRNA. J Biol Chem 1996; 271:20965-72.

42. Yanagisawa K, Tsukamoto T, Takagi T et al. Murine ST2 gene is a mem-

ber of the primary response gene family induced by growth factors. FEBS Lett 1992; 302:51-3.

43. Sims JE, Gayle MA, Slack JL et al. Inetrleukin 1 signaling occurs exclusively via the type I receptor. Proc Natl Acad Sci USA 1993; 90:6155-9.

44. Colotta F, Re F, Muzio M et al. Interleukin-1 type II receptor: a decoy target for IL-1 that is regulated by IL-4. Science 1993; 261:472-5.

45. Muegge K, Williams TM, Kant J et al. Interleukin-1 costimulatory activity on the interleukin-2 promoter via AP-1. Science 1989; 246:249-51.

46. Muegge K, Durum SK. Cytokines and transcription factors. Cytokine 1990; 2:1-8.

47. Muegge K, Vila M, Gusella GL et al. Interleukin 1 induction of the c-jun promoter. Proc Natl Acad Sci USA 1993; 90:7054-8.

48. Bomsztyk K, Sims JE, Stanton TH et al. Evidence for different interleukin 1 receptors in murine B- and T-cell lines. Proc Natl Acad Sci USA 1989; 86:8034-8.

49. Mathias S, Younes A, Kan C-C et al. Activation of the sphingomyelin signal pathway in intact EL4 cells and in a cell-free system by IL-1β. Science 1993; 359:519-22.

50. Cao Z, William JH, Gao X. IRAK: a kinase associated with the interleukin-1 receptor. Science 1996; 271:1128-31.

51. Letsou A, Alexander S, Orth K, et al. Genetic and molecular characterization of tube, a *Drosophila* gene maternally required for embryonic dorsoventral polarity. Proc Natl Acad Sci USA 1991; 88:810-4.

52. Shelton CA, Wasserman SA. Pelle encodes a protein kinase required to establish dorsoventral polarity in the *Drosophila* embryo. Cell 1993; 72:515-25.

53. Luheshi G, Hopkins SJ, Lefeuvre RA et al. Importance of brain IL-1 type II receptors in fever and thermogenesis in the rat. Am J Physiol 1993; 265:E585-91.

54. Takao T, Tracey DE, Mitchell WM et al. Interleukin-1 receptors in mouse brain: Characterization and neuronal localization. Endocrinology 1990; 127:3070-8.

55. Takao T, Newton RC, De Souza EB. Species differences in [^{125}I]interleukin-1 binding in brain, endocrine and immune tissues. Brain Res 1993; 623:172-6.

56. Opp MR, Obal F Jr, Krueger JM. Interleukin-1 alters rat sleep: temporal and dose-related effect. Am J Physiol 1991; 260:R52-8.

57. Hellerstein MK, Meydani SN, Meydani M et al. Interleukin-1-induced anorexia in the rat. J Clin Invest 1989; 84:228-35.

58. Katsuura G, Arimura A, Koves K et al. Involvement of organum vasculosum of lamina terminalis and preoptic area in interleukin 1β-induced ACTH release. Am J Physiol 1990; 258:E163-71.

59. Rivier C, Vale W. Stimulatory effect of interleukin-1 on adrenocorticotropin secretion in the rat: Is it modulated by prostaglandins? Endocrinology 1991; 129:384-8.

60. Dinarello CA, Thompson RC. Blocking IL-1: interleukin-1 receptor an-

tagonist in vivo and in vitro. Immunol Today 1991; 12:404-10.

61. Marquette C, Van Dam AM, Ban E, et al. Rat interleukin-1 beta binding sites in rat hypothalamus and pituitary gland. Neuroendocrinology 1995; 62:362-9.

62. Takao T, Culp SG, Newton RC et al. Type I interleukin-1 (IL-1) receptors in the mouse brain-endocrine-immune axis labelled with 125I-recombinant human IL-1 receptor antagonist. J Neuroimmunol 1992; 41:51-60.

63. Gery I, Gershon RK, Waksman BH. Potentiation of the T-lymphocyte response to mitogens I. The responding cell. J Exp Med 1972; 136:128-42.

64. Munson PJ, Rodbard D. LIGAND: a versatile computerized approach for characterization of ligand-binding systems. Anal Biochem 1980; 297:220-9.

65. Ericsson A, Liu C, Kasckow J et al. Type I interleukin-1 receptor in rat brain: distribution, regulation, and relationship to sites of IL-1 induced cellular activation. J Comp Neurol 1995; 361:681-98.

66. Cunningham ET Jr, Wada E, Carter DB, et al. Localization of interleukin-1 receptor messenger RNA in murine hippocampus. Endocrinology 1991; 128:2666-8.

67. Cunningham ET Jr, Wada E, Carter DB et al. In situ histochemical localization of type I interleukin-1 receptor messenger RNA in the central nervous system, pituitary and adrenal gland of the mouse. J Neurosci 1992; 12:1101-14.

68. Boulant JA, Demieville HN. Responses of thermosensitive preoptic and septal neurons to hippocampal and brain stem stimulation. J Neurophysiol 1977; 40:1356-68.

69. Hori T, Osaka T, Kiyohara T et al. Hippocampal input to preoptic thermosensitive neurons in the rat. Neurosci Lett 1982; 32:155-8.

70. Jacobson L, Sapolsky R. The role of the hippocampus in feedback regulation of the hypothalamic-pituitary-adrenocortical axis. Endocr Rev 1991; 12:118-34.

71. Keller-Wood M, Dallman M. Corticosteroid inhibition of ACTH secretion. Endocr Rev 1984; 5:1-24.

72. Navarra P, Tsagarakis S, Fara MS et al. Interleukin-1 and -6 stimulate the release of corticotropin-releasing hormone-41 from rat hypothalamus in vitro via the eicosanoid cyclooxygenase pathway. Endocrinology 1991; 128:37-44.

73. Tsagarakis S, Gillies G, Rees LH et al. Interleukin-1 directly stimulates the release of corticotropin releasing factor from rat hypothalamus. Neuroendocrinology 1989; 49:98-101.

74. Sharp BM, Matta SG, Peterson PK et al. Tumor necrosis factor-alpha is a potent ACTH secretagogue: comparison to interleukin-1 beta. Endocrinology 1989; 124:3131-5.

75. Yabuuchi K, Minami M, Katsumata S et al. Localization of type I interleukin-1 receptor mRNA in the rat brain. Mol Brain Res 1994; 27:27-36.

76. Katsuura G, Gottschall PE, Arimura A. Identification of a high-affinity receptor for interleukin-1 beta in rat brain. Biochem Biophys Res Commun

1988; 156:61-7.

77. Farrar WM, Kilian PL, Ruff MR et al. Visualization and characterization of interleukin-1 receptors in brain. J Immunol 1987; 139:459-63.

78. Reikerstorfer A, Holz H, Stunnenberg HG et al. Low affinity binding of interleukin-1 beta and intracellular signaling via NF-kappa B identify Fit-1 as a distant member of the interleukin-1 receptor family. J Biol Chem 1995; 270:17645-8.

79. Takao T, Culp SG, De Souza EB. Reciprocal modulation of interleukin-1β (IL-1β) and IL-1 receptors by lipopolysaccharide (endotoxin) treatment in the mouse brain-endocrine-immune axis. Endocrinology 1993; 132:1497-504.

80. Takao T, Nakata H, Tojo C et al. Regulation of interleukin-1 receptors and hypothalamic-pituitary-adrenal axis by lipopolysaccharide treatment in the mouse. Brain Res 1994; 649:265-70.

81. Lieberman AP, Pitha PM, Shin HS et al. Production of tumor necrosis factor and other cytokines by astrocytes stimulated with lipopolysaccharide or a nerotropic virus. Proc Natl Acad Sci USA 1989; 86:6348-52.

82. Michie HR, Spriggs DR, Monogue KR et al. Tumor necrosis factor and endotoxin induce similar metabolic responses in human beings. Surgery 1988; 104:280-6.

83. Butler LD, Layman NK, Riedl PE et al. Neuroendocrine regulation of in vivo cytokine production and effects. I. In vivo regulatory networks involving the neuroendocrine system, interleukin-1 and tumor necrosis factor-α. J Neuroimmunol 1989; 24:143-53.

84. Haour F, Ban E, Milon G et al. Brain interleukin 1 receptors: characterization and modulation after lopopolysacharide injection. Prog Neuro Endocrin Immunol 1990; 3:196-204.

CHAPTER 4

CYTOKINE INVOLVEMENT IN SLEEP RESPONSES TO INFECTION AND PHYSIOLOGICAL SLEEP

James M. Krueger

INTRODUCTION

Sleep is a scientific enigma. The average human will spend 27 years in this state and most will die during it, yet what sleep does for the brain, individual neurons or glia is not defined on any level. It is unlikely that we will ever understand perception, thought, memory or consciousness until our understanding of the regulation and function of sleep improves. Nevertheless, most people intuitively recognize that sleep increases after sleep deprivation or during infection. During the past ten years much evidence has accumulated indicating that sleep responses to infection and physiological sleep are regulated, in part, by humoral mechanisms, and that cytokines such as interleukin-1β (IL-1β) and tumor necrosis factor α (TNF-α) are key elements in this regulation. These studies and the implications they have on sleep function are reviewed herein.

Sleep is defined using physiological measurements such as the electroencephalograph (EEG), the electromyogram (EMG), brain temperature and movement. Sleep is usually divided into two states, non-rapid eye movement sleep (NREMS) and rapid eye movement sleep (REMS), though within the sleep literature subclassifications of each vigilance state are often used. REMS, also called paradoxical sleep, is associated an EEG composed of relatively fast frequency low-voltage waves, muscle relaxation (a flat EMG) interrupted by phasic twitches, and a relatively rapid rise in brain temperature. In contrast, NREMS, also called slow-wave sleep in animals, is characterized by an EEG of high-voltage slow waves (1/2-4 Hz) and regulated drop in brain temperature.[1] The amplitudes of EEG slow waves (also called slow wave activity [SWA]) are thought to be indicative of NREMS intensity, e.g., EEG SWA is greater after sleep-deprivation.[2,3] In most species including man, there is a continuous structured cycling between waking (W), NREMS, and REMS. Individual bouts of NREMS or REMS last only a few minutes in common laboratory species such as rats, mice and rabbits. There is also a strong influence of circadian rhythms on sleep/wake cycle organization, e.g., in rats most sleep occurs during the day whereas in birds and man it occurs primarily during the dark hours (reviewed ref. 3).

A fundamental obstacle confronting sleep research is the failure to define exactly what sleeps. Defining sleep at the single neuron level is impractical since sleep is likely an emergent property of a population of neurons and it is not possible to know if any individual neuron is in a causal pathway leading to sleep. There is much evidence that the entire brain need not sleep (reviewed in ref. 4); sleep in dolphins is the most striking example of this. In dolphins, NREMS only occurs unilaterally at any given time, and unilateral sleep deprivation leads to sleep rebound on the side of the brain deprived, but not on the contralateral side.[5,6] The uncertainty of not knowing the minimal brain unit capable of sleep makes it difficult to know the appropriate level of analysis for mechanistic explanations. Nevertheless, if we accept the tenet that a population of neurons can be in two or more "states," then we can reductionalistically approach the problem of sleep mechanisms. The first step is to define signals that affect populations of neurons (logically, this would also suggest that some levels of analysis, e.g., intracellular signals, would be overly reductionalistic, even though they may be directly affected by sleep). Indeed, much of sleep research has taken the approach of either examining properties of neurons that are widely connected[7] or by characterizing humoral agents that affect those neurons and sleep.[3,8-10] Historically these two approaches developed independently, although they are now viewed as complementary to each other. For example, humoral regulators of sleep likely affect sleep by altering the dynamics of neural circuits (reviewed in refs. 11, 12).

The accumulation of somnogenic substances (SS) in cerebrospinal fluid (CSF) during prolonged W, and the transfer of that CSF and subsequent induction of sleep in recipient animals,[2] provides very strong support for the hypothesis that sleep is regulated, in part, by humoral agents.[12] Indeed many substances affect sleep, however, only a handful of humoral agents are strongly implicated in sleep regulation. The list includes TNF-α, IL-1β, growth hormone-releasing hormone (GHRH), prostaglandin (PG) D_2 (PGD$_2$) and adenosine for

NREMS, and vasoactive intestinal polypeptide (VIP) and prolactin for REMS (reviewed in refs. 2, 8, 9, 13). It is important to recognize that those agents implicated in NREMS affect each other's production and seem to act in concert with each other to affect sleep (reviewed in refs. 8, 12, see Fig. 4.3). Similarly, it is likely that VIP-induction of hypothalamic production of prolactin mRNA[14] and the role of VIP as a hypothalamic releasing factor for pituitary prolactin are involved in REMS regulation (reviewed in ref. 15).

All of the putative SSs thus far identified have other biological activities, some of which are not normally observed during sleep. For example, TNF and some PGs are pyrogenic, yet body temperature decreases upon entry into NREMS.[1,16,17] A model has been constructed (reviewed ref. 12) to illustrate how specific sleep responses may be elicited from multiple pleiotropic SSs. Such models are also applicable to similar problems of specificity confronting the humoral regulation of all physiological functions. Similar issues are also relevant to theories of neuronal regulation of sleep or other physiological processes; e.g., single hypothalamic neurons respond to changes in osmotic pressure, temperature, and glucose concentration.[18] Moreover, neuronal sensitivity to such stimuli is dynamic; e.g., some cells that are insensitive to temperature become more sensitive to heat after exposure to IL-1.[19] Similarly, a single neuron can be part of more than one sensory network.[20] Regardless of such models and considerations, to understand the humoral regulation of sleep, it is first necessary to develop clear evidence implicating a SS in sleep regulation. This review will focus on cytokines in NREMS regulation.

Cytokines are best known, for their roles as inflammatory mediators and in triggering the acute phase response (APR); one facet of the APR is somnolence.[8] The hypothesis that cytokines are involved in physiological processes is less well studied. The central idea of this hypothesis is that, under normal conditions, low basal levels of cytokine production/effects vary in subtle ways with one or more physiological processes. Specificity of response for any one physiological function, e.g., sleep, would arise from a network of interactions of several cytokines/hormones with multiple neuronal groups, each having a different array of sensitivities to different cytokines/hormones (reviewed refs. 12, 13). Removal of any one somnogenic cytokine inhibits normal sleep, alters the cytokine network by changing the cytokine mix, but does not completely disrupt sleep due to the redundant nature of the cytokine network. After pathological disturbances, the production of one or more cytokines would be greatly amplified via microbial (or other pathological) stimuli in a site-specific manner. Such amplified cytokine production would then induce pathology in a manner analogous to the pathologies produced by excessive production of hormones.

SLEEP DURING INFECTION

The APR is an organized host response to microbial challenges that involves many physiological systems. Certain proinflammatory cytokines such a IL-1 and TNF have the capacity to induce the complete array of the APRs. There are several CNS facets of the APR including loss of appetite, fever, social withdrawal, and sleep. Some of these CNS APRs have been extensively studied over many years, e.g., fever. Others such as sleep have only recently been described despite the common experience of feelings of sleepiness and fatigue that accompany infections. Within the past few years much has been learned of how sleep changes during infection and of the causative mechanisms responsible for these sleep responses; these developments are reviewed here.

The initial response of rabbits challenged with bacteria is characterized by excess NREMS. Within a few hours of challenge, rabbits increase the percent of time spent in NREMS from control values of about 45% to 70% or more.[21,22] This period of excess NREMS lasts from 6-24 hours depending upon the route of infectious challenge and bacterial species. In general, NREMS responses to gram-negative bacteria occur with a short latency, but last only a few hours. In contrast, NREMS responses to gram-positive bacteria may take several hours to develop, but then last for 18 or more hours. After the period of excess NREMS, there follows a longer period during which NREMS values are less than corresponding control values. Changes in EEG SWAs are also biphasic and roughly parallel the changes in NREMS. The initial response is one of enhanced EEG SWAs, which as discussed above, could indicate a greater intensity of NREMS. This is followed by periods of EEG SWAs that are less than control values. The phase of NREMS and EEG SWA suppression is also abnormal in the sense that NREMS is more fragmented.

The time course of REMS responses to bacterial challenge is different from that of NREMS responses.[21,22] After challenge, REMS is inhibited within a few hours and remains below control values for two or more days. Similarly, fever develops within a few hours of challenge and often lasts for days. Later in this review the arguments will be presented that sleep responses are not secondary to fever responses.

The sleep and EEG SWA responses induced by bacteria can be blocked by antibiotic treatment.[21] Nevertheless, bacterial replication, per se, is not necessary for the induction of sleep responses. If sufficient doses of dead bacteria are given to rabbits biphasic sleep responses occur although they are shorter in duration than those following viable bacteria inoculation. Further, if bacterial cell wall peptidoglycan is isolated then injected into rabbits, sleep responses occur.[23] If macrophages are fed either bacteria or peptidoglycan they phagocytosize, then digest the cell walls releasing somnogenic muramyl peptides in the process.[24] Muramyl peptides are the monomeric building blocks of cell wall peptidoglycan present in gram-negative and gram-positive bacteria. This process of bacterial digestion and muramyl peptide release could represent a host mechanism for amplifying the response to infectious challenge. Muramyl peptides have been extensively characterized for their ability to induce cytokine production (reviewed in ref. 25), fever[26,27] and sleep[8,27,28] and as immune adjuvants.[29] Structural requirements for muramyl peptide-induced sleep, fever and immune activities are distinct (reviewed in refs. 30-33).

Another mechanism involved in the induction of sleep responses during gram-negative bacterial infection likely involves lipopolysaccharides (LPS) or endotoxins, which are components of gram-negative bacteria cell walls. LPS, as well as its lipid A moiety, are somnogenic.[34,35] They are also well known for their abilities to induce other facets of the APR and cytokine production. There are specific structural requirements for lipid A somnogenic activity, e.g., monophophoric lipid A is less potent than diphosphoric lipid A.[35]

Viral infections, like bacterial infections, are accompanied by the APR including changes in sleep. Several sleep pathologies have been loosely associated with viral infections. The list includes, chronic fatigue syndrome,[36] sudden infant death syndrome,[37] mononucleosis[38] and postviral fatigue syndrome;[39] in none of these cases have direct causative links been established between the viral infections and sleep pathologies. In contrast, there are now several studies in which the direct effects of systemic viral infection on sleep have been determined. In humans HIV infections are associated with excess stage 4 NREMS (stage 4 is the deepest stage of NREMS in humans) during the early asymptomatic stages of HIV infection.[40] This excess stage 4 sleep occurs during the latter half of the night which is very unusual for sleep pathologies. Later on during the course of HIV infection, after AIDS develops, sleep is greatly disrupted.[41] In experimental animals, sleep also changes after viral infections. In mice, influenza viral infections localize to the lungs. Mice with these infections exhibit excess NREMS that lasts for several days. These sleep responses are dependent upon the strain of mouse infected, C57BL/6[42] and Swiss-Webster[43] mice, but not BALB/c,[42] mice exhibit robust NREMS responses to influenza viral infections. Sleep responses also depend upon the strain of virus used; a more virulent strain (H1N1) induced sleep responses of greater magnitude than a less virulent strain (H3N2).[43] In rabbits, influenza virus only undergoes a partial replication thus is an aborted infection. Rabbits inoculated with large doses of viable influenza virus exhibit excess NREMS lasting 4-6 hours. Their excess NREMS is associated with increases in EEG SWAs and with increases in serum anti-viral activity (probably interferon—see below).[44] Unlike what occurs during bacterial infection, rabbit REMS was not altered by influenza virus challenge. An important observation, germane to the mechanisms by which viruses induce sleep responses, is that heat-killed virus was unable to induce sleep effects in rabbits. Further, if rabbits are injected twice with viable virus within 24 hours, they are refractory to the second injection, failing to mount sleep, fever or changes in serum antiviral activity responses.[44,45] Tolerance to influenza virus is also induced by injection of the synthetic double-stranded (ds) RNA polyribo-inosinic:polyribocytidylic (polyI:C) acid 24 hours before virus injection.[45] These observations are consistent with the hypothesis first proposed by Carter and DeClercq[46] that the constitutional symptoms associated with viral diseases are induced by viral dsRNA. There are additional data supporting the idea that viral dsRNA induces sleep responses. Influenza virus dsRNA extracted from infected mouse lungs, but not RNA extracted from control lungs, induces sleep responses in rabbits.[47] Further, a ds 108 mer derived from influenza gene segment 3, but not the corresponding single strands of the 108 mer, induces sleep responses in rabbits.[48] This latter result complements earlier work showing that poly I:C but not poly I or poly C induces sleep responses in rabbits.[49] Finally, like the bacterial products muramyl peptides and LPS, poly I:C and viral dsRNA induce cytokine production, although the array of cytokines induced by each of these stimulants may be distinct, for example, in a macrophage cell line poly I:C, but not LPS, induced IFNα production.[50]

There are also changes in sleep associated with CNS viral infectious.[40,41,51-53] In these cases, however, it is difficult to distinguish between changes in sleep resulting from viral-induced lesions to neural tissues from those induced by viral-induced changes in cytokine production. Nevertheless, CNS viral infections were very important in the development of sleep research. Von Economo[51] showed that viral-induced lesions of the anterior hypothalamus were associated with insomnia. Those findings obtained over 60 years ago, greatly strengthened the view that sleep is an active process which, at the time, was not at all evident.

The most marked changes in sleep in response to infectious challenge are the robust increases in NREMS. However, there are some interesting exceptions to this response. Thus, if rabbits are kept in constant light or constant dark after infectious challenge the magnitudes of NREMS responses are very different. In constant light, NREMS and EEG SWAs responses are greater than those observed in animals kept on a 12-hour light-dark cycle.[54] In contrast, if rabbits are kept in constant dark, NREMS and EEG SWAs responses are substantially less than those observed in animals kept on a 12-hour light-dark cycle. The disease commonly called "sleeping sickness" is caused by the protozoan *Trypanosoma brucei*. In rabbits a related species, *Trypanosoma brucei brucei* (Tbb) induces a chronic disease characterized by periodic parasitemia. During this disease there are periods of excess NREMS and fever which occur during periods of parasitemia.[55,56] However, those periods of excess NREMS are superimposed on a longer term trend of decreased NREMS. Each period of parasitemia is associated with an immune challenge to the host including increased cytokine production.[57-59] These results are thus consistent with the hypothesis that endogenous cytokines are driving sleep responses.

The data reviewed here suggest that sleep responses are one facet of the APR to infectious challenge and that enhanced microbial-induced cytokine production is likely responsible for sleep responses (also see below). Nevertheless, much remains uninvestigated in this area. For example, the fundamental question of whether excess sleep is beneficial to host defense mechanisms remains unanswered (reviewed in ref. 60). Further, rabbits are the only animal in which sleep has been recorded during bacterial infections. In addition, all the infectious challenges given to animals from which sleep has been recorded, have been done at 9:00 a.m. Of great general biological interest are the relationships between the endogenous microbial flora and sleep. Healthy animals[61] and humans[62] treated with antibiotics sleep less than normal. The first somnogenic muramyl peptide described was isolated from human urine and rabbit brain.[63,64] Such findings could suggest that even in normal healthy individuals the endosymbiotic relationship between microbes and host would influence daily sleep. While these speculations are exciting, what is currently clear is that sleep is altered during infectious processes thereby confirming common subjective experience. These changes in sleep associated with pathology likely result from dynamic amplified changes in the brain cytokine network.

INTERLEUKIN-1 (IL-1)

Interleukin-1 is the best characterized humoral agent involved in physiological sleep regulation. The IL-1 family of molecules, IL-1β, IL-1 receptors (IL-1R) Type I and II, the IL-1 receptor antagonist (IL-1ra) the IL-1 accessory protein (IL-1AP) are all constitutively expressed in normal brain (reviewed in ref. 8). IL-1α can also be expressed in the CNS, but whether it is in normal brain remains undecided. IL-1β mRNA, IL-1R mRNA, IL-1ra mRNA and the IL-1AP mRNA all are present in normal brain (see refs. 65-81). Neurons staining for IL-1β-like activity exist in human[73] and rat hypothalamus.[74] IL-1β is also made by glia.[81] There are also several studies showing that IL-1β is upregulated in brain after systemic challenge with LPS.[82-84] Collectively, these studies on the expression of the IL-1 family of molecules lead to the firm conclusion that they are present in normal brain. Several physiological roles for the IL-1 system have been proposed, e.g., regulation of gastrointestinal function,[85] however, the data concerning IL-1 in physiological sleep regulation are the most extensive.

Several laboratories beginning with Moldofsky's in 1986 have described an association of IL-1 plasma levels with sleep-wake cycles.[86-88] In normal people plasma levels of IL-1 are highest at sleep onset[86] and increase during sleep deprivation.[89] Further, if one challenges circulating monocytes with bacteria products in vitro, their ability to produce IL-1 varies with the sleep-wake cycle and is highest at the time of sleep onset.[88] Causative relationships between circulating IL-1 and sleep remain unknown. There are, however, several proposed mechanisms by which systemic cytokines could influence the CNS, e.g., a specific IL-1 transport system from blood to brain,[90] passage of IL-1 into the brain at the OVLT,[91] cytokine stimulation of vagal afferents, and others.[92] Regardless of such possibilities it is likely that brain IL-1 is more relevant to sleep regulation, unfortunately there is currently limited data describing changes in the IL-1 system in brain with sleep. Nevertheless, two reports are encouraging. Thus, Moldofsky's group described changes in IL-1 cerebrospinal fluid levels occurring in phase with the sleep-wake cycle of cats.[93] More recently, hypothalamic IL-1β mRNA was shown to increase after sleep deprivation in rats. There have been no studies linking levels of the IL-1ra, IL-1Rs, IL-1AP or IL-1α to sleep. However, in a strain of knock-out mice lacking the IL-1 type I receptor, the mice fail to respond to IL-1β.[94]

The first study linking IL-1 to sleep was in 1983 in which it was shown that injection of a preparation endogenous pyrogen (containing IL-1) obtained from peritoneal exudate cells induced excess NREMS in rabbits.[95] Subsequently, that work was expanded by the demonstration that various IL-1 preparations, including species-specific recombinant products, induce excess NREMS in rabbits[96] (Fig. 4.1) rats,[97,98] mice[94] and cats,[99] and behavioral signs of sleep in monkeys[100] and sleepiness in humans.[83] Although the general conclusion of these studies is that IL-1β induces increases in NREMS whether given intracerebroventricularly, intravenously or intraperitoneally, its actions on sleep can be complex. For example, in rats, low doses of IL-1β induce increases in NREMS where as high doses inhibit sleep;[97] similar effects are observed in cats.[99] Further, the effects on sleep depend on the time of day IL-1 is administered. Midlevel doses (e.g., 10 ng intracerebroventricularly) of IL-1 given during the night promote NREMS, while the same dose given during the daytime inhibits rat sleep. In rabbits, doses of IL-1 that induce robust increases in NREMS inhibit REMS although, in rats, lower somnogenic doses of IL-1β do not inhibit REMS.[97] IL-1α is also somnogenic, but this is, to date, the only study linking IL-1α to sleep.[101] IL-1β also induces increases in EEG SWAs occurring during sleep.[95,96,101] As indicated above, such supranormal EEG slow waves are thought to be indicative of the intensity of sleep since they also occur after sleep deprivation.

A fragment of human pro-IL-1β corresponding to amino acid residues 208-240 (IL-1$_{208-240}$) also is somnogenic.[101] If rabbits are pretreated with the IL-1ra then given IL-1$_{208-240}$, the somnogenic activity of the fragment is lost thereby suggesting that this fragment interacts with one of the IL-1 receptors.[102] This fragment also induces ICAM-1 expression in a glioblastoma cell line in vitro and this effect is also blocked

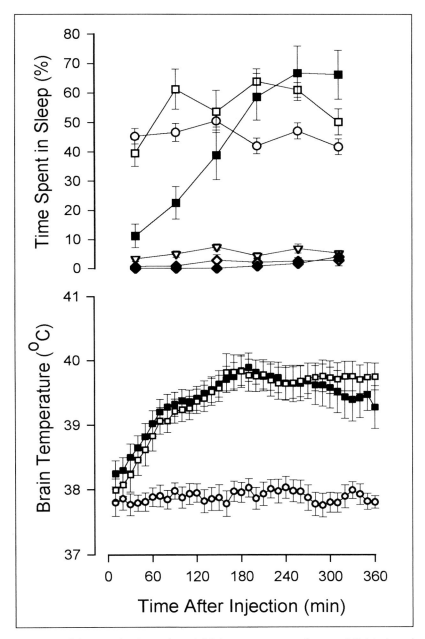

Fig. 4.1. Inhibition of NO synthase inhibits spontaneous sleep and IL-1-induced sleep. Rabbits were injected with either vehicle (○ and ▽), IL-1 alone (□ and ◇) or Nω-nitro-L-arginine methylester (L-NAME) +IL-1 (■ and ◆). After injection EEG, motor activity and brain temperature (Tbr) were recorded for the next 6 hours. L-NAME inhibited sleep and IL-1-induced sleep. In contrast, L-NAME failed to affect IL-1-induced fever. Data reproduced from ref. 169.

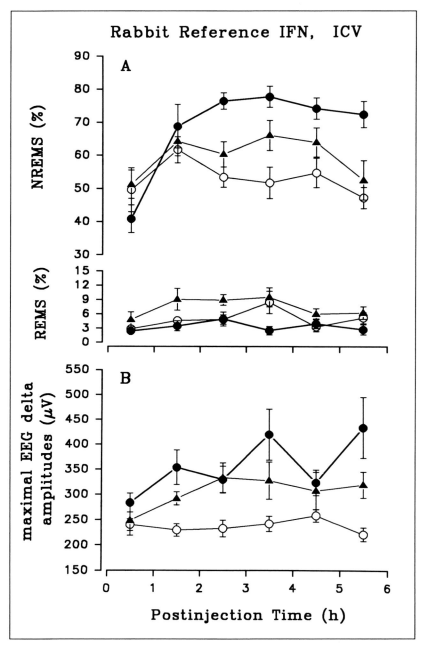

Fig. 4.2. Interferon (IFN) promotes sleep in rabbits. Rabbit reference IFN was injected intracerebroventricularly. Baseline values (O) were obtained after vehicle injection; effects of 25 IU (▲) and 250 IU (●) of rabbit IFN were determined on separate days in the same rabbits. Both doses of IFN enhanced NREMS; only the lower dose enhanced REMS. IFN also enhanced EEG slow wave activity (bottom graph). Similar results can be obtained using other cytokines such as IL-1 (Fig. 4.1), TNF, or acidic FGF (see text). Data reproduced from ref. 161.

by the IL-1ra.[103] Independently, another group reported that residues 208-214 are important for the translocation of IL-1 to the nucleus after receptor-mediated endocytosis of the IL-1-IL-1R complex.[104] The functional manifestations of IL-1 nuclear localization are unknown. Of interest is that the IL-1$_{208-240}$ fragment lacks the ability to stimulate thymocyte proliferation in vitro.[101] In additional sleep studies, the species-specific rabbit IL-1$_{208-240}$ and rat IL-1$_{208-240}$ peptides were shown to be somnogenic in the respective species.[102] This region of IL-1 is highly conserved across species.

The most straightforward data suggesting a role of IL-1 in physiological sleep regulation are those demonstrating that inhibition of IL-1 inhibits sleep in normal animals. Several IL-1 inhibitors have been used in these studies. Anti-IL-1 antibodies reduce spontaneous sleep in rats and rabbits.[105,106] The IL-1ra transiently reduces NREMS for about 1 hour in rabbits,[107] but not rats. A peptide fragment of the soluble IL-1R type I corresponding to amino acids 86-95 (IL-1RF) also inhibits spontaneous NREMS.[108] Further, the IL-1RF[109] and anti-IL-1 antibodies[105] reduce sleep rebound after sleep deprivation, thereby suggesting that IL-1 is responsible, in part, for sleep recovery after sleep deprivation. Furthermore, the soluble IL-1R type I[110] and the IL-1RF[108] attenuate MDP-induced NREMS responses thus providing indirect evidence that IL-1 is also involved in sleep responses to infections.

Other inhibitors of IL-1 also inhibit sleep (Fig. 4.3). Alpha melanocyte stimulating hormone (αMSH) inhibits many of the actions of IL-1 including IL-1-induction of NREMS;[111] αMSH also reduces spontaneous sleep. Similarly, PGE$_2$,[9,112] corticotropin releasing hormone (CRH)[113] and IL-10[114] all inhibit IL-1 production and spontaneous NREMS. Finally, some of the somnogenic actions of IL-1 could be mediated via nitric oxide; inhibition of NO synthase blocks IL-1-enhanced NREMS (Fig. 4.1) although IL-1-induced fevers are not affected (relationships between thermoregulation and sleep are discussed below).

In summary, the presence of the IL-1 system in brain, IL-1 induction of NREMS, and inhibition of NREMS by IL-1 inhibitors collectively support the hypothesis that IL-1 is a key substance in physiological sleep regulation.

TUMOR NECROSIS FACTOR α (TNF-α)

TNF, like IL-1, may be involved in physiological sleep regulation as well as in sleep responses to pathologies and sleep deprivation. The TNF family of molecules is found in normal brain. TNF-α mRNA and TNF receptor mRNA are expressed in normal brain.[115-117] TNF is produced by astrocytes[118] and TNF-α immunoreactive neurons are found in the hypothalamus,[115] an area involved in NREMS regulation. The soluble TNF receptor is a normal constituent of human CSF.[119] TNF seems to be important during brain development.[120-122] Further,

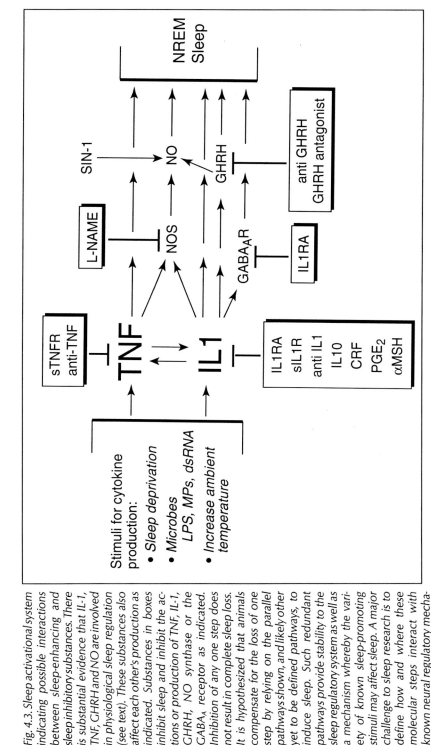

Fig. 4.3. Sleep activational system indicating possible interactions between sleep-enhancing and sleep inhibitory substances. There is substantial evidence that IL-1, TNF, GHRH and NO are involved in physiological sleep regulation (see text). These substances also affect each other's production as indicated. Substances in boxes inhibit sleep and inhibit the actions or production of TNF, IL-1, GHRH, NO synthase or the GABA_A receptor as indicated. Inhibition of any one step does not result in complete sleep loss. It is hypothesized that animals compensate for the loss of one step by relying on the parallel pathways shown, and likely other yet to be defined pathways, to induce sleep. Such redundant pathways provide stability to the sleep regulatory system as well as a mechanism whereby the variety of known sleep-promoting stimuli may affect sleep. A major challenge to sleep research is to define how and where these molecular steps interact with known neural regulatory mechanisms; e.g., one exciting possibility is that the IL-1-GABA_A interaction could occur in thalamo-cortical circuits thereby enhancing EEG synchronization. (→) indicates stimulation; (⊣) indicates inhibition; (→→→) indicates additional steps likely involved in sleep regulation.

pathological stimuli such as LPS or neurotropic virus induce enhanced brain TNF production.[117] TNF also has direct effects on hypothalamic neurons[123-125] suggesting that it could be involved in neuroregulatory processes. An important indirect link to sleep regulation is the association between TNF-α and histaminergic neurons.[126] Antihistamines are widely used as somnogenic agents. Neurotoxic lesions of histaminergic neurons decrease TNF-α in the hypothalamus while enhancing TNF-α production in the hippocampus.[127] Electrolytic lesions of the anterior hypothalamus, which would include lesion of histaminergic neurons, result in insomnia.[128] These data suggest that neuronal histamine is involved, in part, in the regulation of the brain TNF-α system. Data from other cells suggest that histamine inhibits TNF-α gene expression and TNF synthesis (e.g., ref. 129).

Intracerebroventricular or intravenous injection of exogenous TNF-α enhances NREMS in rabbits,[96,130,131] rats[132] and mice.[94] The TNF-induced NREMS is evident within the first postinjection hour then continues 12 or more hours. Doses of TNF that elicit pronounced increases in NREMS also induce increases in EEG SWAs, fever and decreases in REMS in rabbits. Lower somnogenic doses do not affect REMS in rabbits[96] and mice.[94] Administration of exogenous TNF-β[133] or peptide fragments of TNF-α or TNF-β also enhance NREMS in rabbits.[134] Evidence for the involvement of TNF in physiological sleep regulation comes from experiments in which inhibitors were given to normal animals. Anti-TNF antibodies suppress spontaneous sleep in rats and rabbits.[135] The soluble TNF receptor (also called the TNF binding protein)[130] or a peptide fragment of the soluble receptor (TNFRF) also inhibit spontaneous NREMS in rabbits.[131] The TNF 55 kDa receptor is involved in TNF somnogenic activity since a knock-out strain of mice lacking this receptor fails to mount NREMS responses if given TNF-α.[94] The TNFRF also attenuates sleep rebound after sleep deprivation as well as MDP-enhanced NREMS.[136] Further, the TNFRF blocks the enhancement of NREMS associated with mild increases in ambient temperature.[137]

There is also indirect evidence from humans supporting a role for TNF in sleep regulation. In normal people, TNF plasma levels vary in phase with EEG SWA.[138] Sleep deprivation, but not other stressors, prime monocytes for LPS-induced TNF production.[139] Others have also shown that sleep deprivation increases LPS-[140] or streptococcal-induced[88] TNF production. Although there is a specific TNF transport system across the blood-brain barrier, the respective roles of circulating vs. central TNF in sleep regulation remain unknown.

The relationship between TNF and IL-1 in sleep regulation are only beginning to be investigated. IL-1 and TNF induce each other's production.[141] Their somnogenic actions seemed to be linked to each other (e.g., Fig. 4.3), yet there is also evidence that they can independently affect sleep. In rabbits treated with IL-1, the IL-1-induced NREMS responses can be attenuated if animals are pretreated with the TNFRF.[142] Conversely, TNF-induced NREMS responses are attenuated in rabbits pretreated with the IL-1RF.[142] In contrast, knock-out mice lacking the IL-1 type I receptor respond to TNF-α by increasing NREMS and other knock-out mice, lacking the TNF 55 kDa receptor, respond to IL-1.[94] These latter data from the knock-out strains suggest that these cytokines can, under appropriate conditions, affect sleep independently of each other.

ACIDIC FIBROBLAST GROWTH FACTOR

Fibroblast growth factors (FGFs) belong to a growth factor family that includes IL-1α and IL-1β.[143,144] Two forms of FGF, acidic FGF (aFGF), and basic FGF (bFGF) share a limited amino acid homology with IL-1 and they have three-dimensional topologies similar to IL-1. Further, like IL-1, they lack a hydrophilic NH_2-terminal lead sequence.[145] FGFs also share several biological activities with IL-1,[146] e.g., they inhibit food intake.[147] FGFs and their receptors are also found in the brain. Intracerebroventricular administration of recombinant human FGF induces dose-related increases in NREMS in rabbits[148] and rats.[149] In rabbits, the lowest effective dose is 0.1 μg/rabbit; this dose is about 20-fold the minimum effective dose of IL-1β in rabbits. Acidic FGF-induced increases in NREMS are evident within the first hour of its injection. Somnogenic doses of aFGF also inhibit REMS and at the highest somnogenic dose tested aFGF is also pyrogenic after a 2-hour delay. These activities are lost if aFGF is heat-inactivated before injections.[148] Similar doses of bFGF lack any of these activities.[148] The reasons behind the differential effects of aFGF and bFGF on sleep are not understood. Both FGFs share many activities and compete with each other for receptors. They share about a 55% amino acid homology and have similar three dimensional structures. Nevertheless, in addition to their effects on sleep, other differences in the biological actions of aFGF and bFGF have been reported. For example, bFGF inhibits aspartate aminotransferase activity whereas aFGF stimulates it.[150] Brain FGF is a better stimulator of glutamate decarboxylase than is aFGF.[150] Acidic FGF and bFGF also have differential effects on dopamine uptake by neurons and glia[151] and on nerve growth factor secretion by astrocytes.[152] While all these substances differentially affected by aFGF and bFGF are likely involved in some aspect of sleep regulation, it is premature to speculate as to their involvement in the differential effects of aFGF and bFGF on sleep. Nevertheless, these observations offer promising avenues to investigate cytokines in sleep regulation.

INTERFERONS (IFNs)

IFNs were initially characterized by their ability to inhibit viral proliferation. Most nucleated cells can produce one or more forms of IFNα and IFNβ in response to infectious agents[153] and IFNα receptors are in brain.[154] Indeed, it is likely that IFNs are intimately involved in sleep responses that occur during viral infections. The somnogenic properties of IFNs were first suggested in studies of patients undergoing IFN therapy; the patients reported excessive sleepiness.[155] Several subsequent studies also implicate IFNs in sleep responses occurring during infection though, to date, there are no data suggesting that IFNs are involved in physiological sleep regulation. For example, serum antiviral activity (likely IFN) increases in temporal correlation with enhanced sleep and fever in rabbits inoculated with influenza virus.[44] Administration of exogenous IFNs also suggests that they affect sleep. Thus, human IFNα enhances cortical EEG synchronization in rats,[156-158] reduces latency to REMS in monkeys[159] and promotes NREMS, but not REMS, in rabbits whether given intracerebroventricularly or intravenously,[160] although the doses used to elicit these effects are very high. In one study, a species specific IFN was used in rabbits (Fig. 4.2); the dose required to induce enhanced NREMS responses were much lower than doses of human-IFNs needed to promote sleep in rabbits.[161] In this latter study, IFNβ also promoted sleep; however, its somnogenic activity could not be heat inactivated thereby suggesting that it may have been contaminated with microbial products[162] since the IFN used was produced by recombinant methods. IFNγ has, to our knowledge, not been tested for somnogenic activity although large doses of IFNγ in cancer patients are associated with a flu-like syndrome.[163] Microinjection of either human recombinant IFNα or rat IFN into the locus ceruleus (an area involved in sleep regulation) of rats induces enhancements of sleep and EEG SWA.[157] IFN can induce IL-1 activity[164] and it is thus possible that the somnogenic actions of IFN may be mediated via IL-1.

OTHER CYTOKINES IL-2, IL-6 AND IL-10

Interleukin-2 (IL-2) has also been tied to sleep, however convincing data for its role in sleep regulation are lacking. IL-1 has the capacity to induce IL-2 production and this suggests the possibility that IL-1 somnogenic actions could be mediated via IL-2. Further, IL-2 and its receptors are produced in the brain[165,166] and IL-2 can enter the brain by a nonspecific transport system. Moldofsky and colleagues[86] were the first to show that plasma IL-2-like activity is greater during sleep than during daytime wakefulness in humans. They confirmed that observation in a later study and extended it by showing that sleep deprivation induces an even greater increase in plasma IL-2 levels during subsequent sleep.[89] Subsequently, Uthgenannt et al[88] also reported enhanced IL-2 levels during sleep. DeSarro and colleagues tested the ability of IL-2 to induce sleep in rats.[157] If IL-2 is injected into the third ventricle or microinjected into the locus ceruleus it enhances sleep. In contrast, if injected into the hippocampus or ventral medial hypothalamus it induces increased locomotor activity. In another study, human recombinant IL-2 failed to affect rabbit sleep, however, rat IL-2 did induce excess NREMS in rats.[167] That report also noted that the rat IL-2 preparation could not be heat inactivated thereby leaving open the possibility that microbial contaminants are responsible for the somnogenic actions observed.[162] IL-2 could thus be involved in sleep regulation, but clean studies with proper controls are needed.

Interleukin-6 (IL-6) serum levels may also vary in humans with the sleep-wake cycle.[140] However, in another study, human IL-6 failed to affect sleep in rabbits although the doses tested did induce fevers.[168] Nevertheless, since species specific IL-6 was not used, this study should be repeated using species-specific IL-6 preparations. Interleukin-10 (IL-10) is an anti inflammatory cytokine. IL-10 inhibits synthesis of many pro-inflammatory cytokines including IL-1 and TNF. IL-10 also inhibits NREMS in rats.[114] These latter results were obtained from normal rats, and thus support the hypothesis that some pro-inflammatory cytokines are involved in physiological sleep regulation.

THERMOREGULATION AND SLEEP

All somnogenic cytokines are also pyrogenic if sufficient doses are used. It is thus logical to ask whether their somnogenic actions are secondary to febrile responses. The answer is clearly no. IL-6 is pyrogenic, but not somnogenic.[168] The pyrogenic actions of IL-1 can be blocked with antipyretics without affecting sleep responses.[95] Conversely, inhibition of NO synthase blocks IL-1-induced sleep responses but not fever.[169] Low doses of IL-1 induce sleep responses, but not fever in rats.[97] In addition, several other substances are pyrogenic yet inhibit sleep, e.g., CRH.[113] Others are cryogenic and inhibit sleep[111] (e.g., αMSH). In rabbits, MDP is somnogenic and pyrogenic during the day; at night, it is more somnogenic than pyrogenic.[170] The time courses of sleep and fever responses during infection are distinct.[21]

In contrast to fever, another aspect of thermoregulation is tightly coupled to sleep. There are small changes in brain temperature that occur with different vigilance states. For example, the transition between W and NREMS is associated with a regulated decrease in brain temperature.[1] In contrast, upon entry into REMS there is a rapid rise of 0.1-0.3°C of brain temperature. These sleep-coupled temperature changes persist throughout the circadian rhythm of temperature[17] and during fever.[16] Somnogenic cytokines fail to affect these sleep-coupled temperature changes.

Other sleep-thermoregulatory relationships may be cytokine-dependent. For example, acute mild increases in ambient temperature are associated with increases in sleep. These ambient temperature-induced sleep responses are blocked if rabbits are pretreated with a soluble TNF receptor.[137] Sleep, body temperature, and plasma levels of IL-1 vary over estrous/menstrual cycles.[171] Similarly, exercise is associated with body temperature changes, excess subsequent sleep and changes in IL-1 plasma levels.[172]

MECHANISMS OF CYTOKINE INDUCED SLEEP

IL-1 and TNF affect the production of other humoral agents implicated in sleep regulation and also directly affect neurons; both mechanisms are likely involved in their somnogenic actions. The growth hormone (GH)/GHRH/somatostatin (SRIF) axis is affected by IL-1 and TNF and it is implicated in sleep regulation. GH release is coupled to the onset of stage 4 sleep in humans.[173,174] Both GH and SRIF enhance REMS (reviewed in ref. 13) while GHRH enhances NREMS and REMS in rats[175-177] and rabbits.[178] GHRH is also an effective somnogen in humans.[179,180] Inhibition of GHRH using either antibodies to GHRH[181] or a GHRH-peptide antagonist[182] inhibits sleep. Anti-GHRH also inhibits sleep rebound after sleep deprivation.[181] TNF and IL-1 induce GH release or production.[183-185] The effect of IL-1 on GH release is mediated via the hypothalamus[184,186] and this likely involves GHRH because anti-GHRH blocks IL-1-induced GH release.[187] In rats, there is additional evidence linking IL-1 somnogenic actions to the GH/GHRH axis. Thus, in rats, low doses of IL-1 promote NREMS and GH release whereas higher doses inhibit sleep and GH release.[97,184] Finally, the somnogenic actions of IL-1 are inhibited by anti-GHRH antibodies.[188] Collectively these data strongly suggest that GHRH is involved in physiological sleep regulation and that IL-1 elicits its affects on sleep, in part, via GHRH.

Other hormonal systems are also likely involved in the somnogenic actions of cytokines. For example, IL-1 induces release of CRH. CRH induces pituitary release of adrenocorticotrophic hormone (ACTH) that in turn stimulates adrenal production of glucocorticoids. Glucocorticoids feedback to inhibit IL-1 production and enhance IL-1 receptor production.[189,190] CRH, ACTH and glucocorticoids all inhibit sleep (reviewed in refs. 12, 191). CRH also inhibits IL-1-enhanced NREMS.[113] It is likely that the CRH/ACTH/glucocorticoid axis operates as a negative feedback system in sleep regulation perhaps being responsible for the observations that high doses of IL-1 inhibit, rather than promote, sleep.

Several neurotransmitter systems are either altered by or affect cytokines; many of these are likely to be linked to the actions of cytokines on sleep. For example, IL-1 activates hypothalamic adrenergic, dopaminergic and serotonergic systems.[192-198] TNF, IL-1 and FGF all modulate Ca^{++} currents in neurons.[199,200] Serotonin has the capacity to induce IL-1 production in vitro.[201] FGF and IL-1 affect GABAergic mechanisms. This latter action of IL-1 can be directly linked to sleep. IL-1 induces, via the IL-1 receptor and the $GABA_A$ receptor, increased Cl^- permeability;[202] the IL-1ra antagonizes this action. The net effect of IL-1 is hyperpolarization of GABA receptive neurons. Steriade and colleagues[7] have demonstrated that hyperpolarization of GABA receptive neurons in thalamo-cortical circuits is responsible for EEG synchronization, which of course is a cardinal sign of NREMS.

Another neurotransmitter system affected by cytokines is nitric oxide (NO). Many cytokines including IL-1 and TNF (reviewed in ref. 203)

as well as some hormones, e.g., GHRH,[204] can enhance NO production. Inhibition of NO synthase using arginine analogs inhibits spontaneous sleep and IL-1-induced sleep in rats[205,206] and rabbits.[169] Further, administration of NO donors, substances that decompose releasing
NO such as S-nitroso-N-acetylpenicillamine, induce prolongation of
sleep.[207] Finally, NO synthase activity has a relatively high amplitude
of its circadian rhythm in the hypothalamus.[208]

To affect sleep, cytokines probably change either directly or indirectly neuronal firing rates. Indeed several cytokines have been extensively studied in this regard (reviewed in refs. 209-211). For example,
IL-1 inhibits long-term potentiation in hippocampal neurons[212,213] and
affects hypothalamic neural activity (for example, see ref. 209). Much
of this type of work was done within the context of temperature and
feeding regulation. However, it may also be relevant to sleep regulation since the areas in brain that have been the focus of this work,
e.g., hypothalamus and hippocampus, are also involved in sleep regulation. Further, changes in vigilance state affect firing patterns and/or
responsiveness of most neurons. Nevertheless, since sleep is an emergent property of populations of neurons and since at a single neuron
level it is not possible to know if it is in a causal pathway leading to
sleep, interpretation of studies focusing on cytokine actions on single
neurons is difficult within the context of sleep regulation.

HUMORAL MECHANISMS OF SLEEP AND SLEEP FUNCTION

The efficacy of synaptic transmission is variable and dependent, in
part, on how often the synapse is used (reviewed in ref. 214). Synaptic plasticity and subsequent changes in neural circuit dynamics are
posited to play a role in learning, memory, and motor function.[215]
These use-dependent changes in the microcircuitry necessarily imply
that changes in electrical potential across the neuron membrane some-
how eventuate in the production of synaptic structural molecules such
as neuronal cell adhesion molecules (NCAM). Neural depolarization
is associated with the release or uptake of many molecules including
K^+, neurotransmitters, adenosine, Ca^{++} etc. These substances likely affect nearby cells and indeed some have been shown to enhance glia
production of cytokines.[201] The released cytokines would in turn have
two paracrine/autocrine actions; one involved in the neural mechanisms of sleep and the other with a synaptic restructuring process.
Cytokines directly affect neurons (see above) and these actions likely
trigger the neural circuitry events involved in sleep generation. For
example, IL-1 augments $GABA_A$ receptor function; if this operates within
thalamocortical circuits it would lead to greater hyperpolarization and
subsequent EEG synchronization.[7,216] Cytokines also are capable of
inducing CAM production (103) including NCAMS; this likely occurs in the same cells whose membranes potentials were altered by

cytokines. Previously, we developed the argument that the primordial function of sleep is to maintain a synaptic superstructure that is necessary for the animals' long-term adaptability to environmental challenge, yet is insufficiently stimulated during W to be maintained.[4,11] Cytokine involvement in membrane (sleep mechanism) as well as nuclear events such as altered gene expression (CAM) (sleep function) suggest that sleep mechanism and function can not be separated at the cellular level. Nevertheless, since single neurons often have thousands of synapses, the manifestations of the cellular function of sleep also occur within populations of neurons. In fact, it is reasonable to argue that single cells do not sleep even though sleep function is expressed at that level.

Regardless of such considerations it is currently clear that certain cytokines are intimately involved in sleep regulation. Those findings coupled with our extensive knowledge of the cellular actions of cytokines have greatly improved our understanding of sleep.

ACKNOWLEDGMENTS

This work was supported in part by grants from the National Institutes of Health (NS25378, NS27250 and NS31453) and the Office of Naval Research (N00014-90-J-1069). I thank Dr. Levente Kapás for reviewing this manuscript and Drs. Fang and Takahashi for allowing me to mention their results not yet published in full form.

REFERENCES

1. Heller HC, Glotzbach SF. Thermoregulation during sleep and hibernation. Int Rev Physiol 1977; 15:147-187.
2. Pappenheimer JR, Koski G, Fencl V et al. Extraction of sleep-promoting factor S from cerebrospinal fluid and from brains of sleep-deprived animals, J Neurophysiol 1975; 38:1299-1311.
3. Borbély AA, Tobler I. Endogenous sleep-promoting substances and sleep regulation. Physiol Rev 1989; 69:605-670.
4. Krueger JM, Obál F Jr. A neuronal group theory of sleep function. J Sleep Res 1993; 2:63-69.
5. Mukhametov LM. Sleep in marine mammals. Exp Brain Res 1984; Suppl 8:157-172.
6. Oleksenko AI, Mukhametov LM, Polyakova IG et al. Unihemispheric sleep deprivation in bottle nose dolphins. J Sleep Res 1992; 1:40-44.
7. Steriade M, McCarley RN. Brainstem Control of Wakefulness and Sleep. New York: Plenum Press, 1990.
8. Krueger JM, Majde JA. Microbial products and cytokines in sleep and fever regulation. Crit Rev in Immunology 1994; 14:355-379.
9. Hayaishi O. Sleep-wake regulation by prostaglandin D_2 and E_2. J Biol Chem 1988; 263:14593-14597.
10. Krueger JM, Obál F Jr. Growth hormone releasing hormone and interleukin-1 in sleep regulation. FASEB J 1993; 7:645-652.

11. Krueger JM, Obál F Jr, Kapás L et al. Brain organization and sleep function. Beh Brain Res 1995; 177-185.

12. Krueger JM, Obál F Jr, Opp M et al. Somnogenic cytokines and models concerning their effects on sleep. Yale J Biol Med 1990; 63:157-172.

13. Krueger JM, Obál F Jr, Opp M et al. Growth hormone-releasing hormone and interleukin-1 in sleep regulation. FASEB J 1993; 7:645-652.

14. Bredow S, Kacsóh B, Obál F Jr et al. Increase of prolactin mRNA in the rat hypothalamus after intracerebroventricular injection of VIP or PACAP. Brain Res 1994; 660:301-308.

15. Roky R, Obál F Jr, Valatx JL et al. Prolactin and rapid eye movement. Sleep 1995; 18:536-542.

16. Walter J, Davenne D, Shoham S et al. Brain Temperature changes coupled to sleep state persist during interleukin 1-enhanced sleep. Am J Physiol 1986; 250:R96-R103.

17. Obál, F Jr, Rubicsek G, Alföldi P et al. Changes in the brain and core temperatures in relation to the various arousal states in rats in the light and dark periods of the day. Pflügers Arch 1985;404:73-79.

18. Silva NL, Boulant JA. Effects of osmotic pressure, glucose, and temperature on neurons in preoptic tissue slices. Am J Physiol 1984; 247:R335-45.

19. Eisenman JS. Electrophysiology of the anterior hypothalamus: Thermoregulation and fever. In: Milton J, ed. Cell Pyretics and Antipyretics. Berlin: Springer-Verlag, 1982:187-217.

20. Hooper SL, Moulins M. Swithching of a neuron from one network to another by sensory-induced changes in membrane properties. Science 1989; 244:1587-1589.

21. Toth LA, Krueger, JM. Alteration of sleep in rabbits by Staphylococcus aureus infection. Infect Immun 1988; 56:1785-1791.

22. Toth LA, Krueger JM. Effects of microbial challenge on sleep in rabbits. FASEB J 1989; 3:2062-2066.

23. Johannsen L, Toth LA, Rosenthal RS et al. Somnogenic, pyrogenic and hematologic effects of bacterial peptidoglycan. Am J Physiol 1990; 259:R182-R186.

24. Johannsen L, Wecke J, Obál F Jr et al. Macrophages produce somnogenic and pyrogenic muramyl peptides during digestion of staphylococci. Am J Physiol 1991; 260:R126-R133.

25. Dinarello CA, Krueger JM. Induction of interleukin-1 by synthetic and naturally occurring muramyl peptides. Fed Proc 1986; 45:2545-2548.

26. Kotani S, Watanabe Y, Shimono T et al. Correlation between the immunoadjuvant activity and pyrogencities of synthetic N-acetylmuramyl-peptides or -amino acids. Biken J 1976; 19:9-13.

27. Masek K. Immunopharmacology of muramyl peptides. Fed Proc 1986; 45:2549-2551.

28. Kadlecová O, Masek K. Muramyl dipeptide and sleep in rats. Meth Find Exp Clin Pharmacol 1986; 8:111-115.

29. Chedid L. Immunopharmacology of muramyl peptides: New Horizons Prog Immunol 1983; 5:1349-1358.

30. Krueger JM, Johannsen L. Bacterial products, cytokines and sleep. In: Lernmark A, Dyrberg T, Terenius T, Hökfelt B ed(s). Molecular Mimicry in Health and Disease. Amsterdam: Elsevier, 1988: 35-46.

31. Krueger JM, Walter J, Karnovsky ML et al. Muramyl peptides: Variation of somnogenic activity with structure. J Exp Med 1984; 159:68-76.

32. Krueger JM, Karnovsky ML, Martin SA et al. Peptidoglycans as promoters of slow-wave sleep. II. Somnogenic and pyrogenic activities of some naturally occurring muramyl peptides; correlations with mass spectrometric structure determination. J Biol Chem 1984; 259:12659-12662.

33. Krueger JM, Rosenthal RS, Martin SA et al. Bacterial peptidoglycan as modulators in sleep. I. Anhydro forms of muramyl peptides enhance somnogenic potency. Brain Res 1987; 403:249-266.

34. Krueger JM, Kubillus S, Shoham S et al. Enhancement of slow-wave sleep by endotoxin and lipid A. Am J Physiol 1986; 251:R591-R597.

35. Cady AB, Kotani S, Shiba T et al. Somnogenic activities of synthetic lipid A. Infect. Immunity 1989; 57:396-403.

36. Komaroff AL. Chronic fatigue syndromes: relationships to chronic viral infections. J Virol Meth 1988; 21:3-10.

37. Hofmann HJ, Damus K, Hillman L et al. Risk factors for SIDS: results of the National Institute of Child Health and Human Development SIDS cooperative epidemiological study. Ann N Y Acad Sci 1988; 533:13-30.

38. Guillenminault C, Mondini S. Mononucleosis and chronic daytime sleepiness: a long-term follow-up study. Arch Intern Med 1986; 146:1333-1335.

39. Behan PO, Behan WM, Gow JW et al. Enteroviruses and postviral fatigue syndrome. Ciba Found Symp 1993; 173:146-154.

40. Norman SE, Chediak HD, Kiel M et al. Sleep disturbances in HIV-infected homosexual men. AIDS 1990; 4:775-781.

41. St. Kubicki H, Henkes H, Terstegge K, Ruf B. AIDS-related sleep disturbances-a preliminary report. In: St. Kubicki H, Henkes H, Bienzle K, Pokle HD eds. HIV and Nervous System. Stuttgart: Gustav-Fisher, 1988; 97-105.

42. Toth LA, Reha JE, Webster RG. Strain differences in sleep and other pathophysiological sequelae of influenza virus infection in naive and immunized mice. Neuroimmunol 1995; 58:89-99.

43. Fang J, Sanborn CK, Renegar KB et al. Influenza viral infections enhance sleep and weight loss in mice. Proc Soc Exptl Med Biol 1995; 210:242-52.

44. Kimura-Takeuchi M, Majde JA, Toth LA et al. Influenza virus-induced changes in rabbit sleep and acute phase responses. Am J Physiol (Regulatory Integrative Comp. Physiol. 32) 1992; 263:R1115-R1121.

45. Kimura-Takeuchi M, Majde JA, Toth LA et al. The role of double-stranded RNA in the induction of the acute-phase response in an abortive influenza virus infection model. J Infect Dis 1992; 166:1266-1275.

46. Carter WA, De Clercq E. Viral infection and host defense. Science 1974; 186:1172-1178.

47. Majde JA, Brown RK, Jones MW et al. Detection of toxic viral-associated double-stranded RNA (dsRNA) in influenza-infected lung. Microbial

Pathogenesis 1991; 10:105-115.

48. Bredow S, Fang J, Guha-Thakurta N et al. Synthesis of an influenza double-stranded RNA-oligomer that induces fever and sleep in rabbits. Sleep Res 1995; 24A:101.

49. Krueger JM, Majde JA, Blatteis CM et al. Polyriboinosinic:polyribocytidylic acid (Poly I:C) enhances rabbit slow-wave sleep. Am J Physiol 1988; 255:R748-R755.

50. Kimura M, Toth LA, Agostini H et al. Comparison of acute phase responses induced in rabbits by lipopolysaccharide and double-stranded RNA. Am J Physiol 1994; 67:R1596-R1603.

51. Von Economo C. Sleep as a problem of localization. J Nerv Ment Dis 1930; 71:249-259.

52. Gourmelon P, Briet D, Clarencon D et al. Sleep alterations in experimental street rabies virus infection occur in the absence of major EEG abnormalities. Brain Res 1991; 554:159-165.

53. Gourmelon P, Briet D, Court L et al. Electrophysiological and sleep alterations in experimental mouse rabies. Brain Res 1986; 398:128-140.

54. Toth LA, Krueger JM. Lighting conditions alter Candida albicans-induced sleep responses. Am J Physiol 1995; 269:R1441-R1447.

55. Toth LA, Tolley EA, Broady R et al. Sleep during experimental trypanosomiasis in rabbits. PSEBM 1994; 205:174-181.

56. Buguet A, Bent J, Tapie P et al. Sleep-wake cycle in human African trypasomiasis. J Clin Neurophysiol 1993; 10:190-196.

57. Rouzer CA, Cerami A. Hypertriglyceridemia associated with Trypanosoma brucei brucei infection in rabbits: Role of defective triglyceride removal, Molec Biochem Parasitol 1980; 2:31-38.

58. Askonas BA, Bancroft GJ. Interaction of African trypanosomes with the immuned system, Phil Trans R Soc Lond 1984; 307:41-50.

59. Bancroft GJ, Sutton CJ, Morris AG et al. Production of interferons during experimental African trypanosomiasis. Clin Exp Immunol 1983; 52:135-143.

60. Toth LA, Tolly EA, Krueger JM. Sleep as a prognostic indicator during infectious disease in rabbits, Proc Soc Exptl Biol Med 1993; 203:179-192.

61. Brown R, Price RJ, King MG et al. Are antibiotic effects on sleep behavior in the rat due to modulation of gut bacteria? Physiol and Behav 1990; 48:561-565.

62. Rhee YH, Kim HI. The correlation between sleeping-time and numerical range in intestinal normal flora in psychiatric insomnia patients. Bull Nat Sci Chungbuk Natl Univ 1987; 1:159-172.

63. Krueger JM, Pappenheimer JR, Karnovsky ML. Sleep-promoting effects of muramyl peptides. Proc Natl Acad Sci USA 1982; 79:6102-6106.

64. Krueger JM, Pappenheimer JR, Karnovsky ML. The composition of sleep-promoting factor isolated from human urine. J Biol Chem 1982; 257:1664-1669.

65. Farrar WL, Hill JM, Harel-Bellan A et al. The immune logical brain. Immunol Rev 1987; 100:361-377.

66. Berkenbosch F, Robakis N, Blum M. Interleukin-1 in the central nervous system: A role in the acute phase response and in brain injury, brain development and the pathogenesis of Alzheimer's disease, In: Frederickson RCA, McGaugh JC, Felten DL ed(s). Peripheral Signaling of the Brain. Toronto: Hogrefe & Hunter Pub, 1991:131-145.

67. Higgins GA, Olschowka, JA. Induction of interleukin-1β mRNA in adult rat brain. Mol Brain Res 1991; 9:143-148.

68. Bandtlow CE, Meyer M, Lindholm D et al. Regional and cellular codistribution of interleukin-1β and nerve growth factor mRNA in the adult rat brain: Possible relationship to the regulation of nerve growth factor synthesis. J Cell Biol 1990; 111:1701-1711.

69. Yan HQ, Banos MA, Herregodts P et al. Expression of interleukin-1 (IL-1) beta, IL-6 and their respective receptors in the normal rat brain and after injury. Eur J Immunol 1992; 22:2963-2971.

70. Minami M, Kuraishi Y, Yamaguchi T et al. Convulsants induce interleukin-1β messenger RNA in rat brain. Biochem Res Comm 1990; 171:832-837.

71. Cunningham ET Jr, de Souza EB. Interleukin 1 receptors in the brain and endocrine tissues. Immunology Today 1993; 14:171-176.

72. Farrar WL, Kilan PL, Ruff MR et al. Visualization and characterization of interleukin-1 receptors in brain. J Immunol 1987; 139:459-463.

73. Breder CD, Dinarello CA, Saper CB. Interleukin-1 immunoreactive innervation of the human hypothalamus. Science 1988; 240:321-324.

74. Lechan RM, Toni R, Clark BD et al. Immunoreactive interleukin-1β localization in the rat forebrain. Brain Res 1990; 514:135-140.

75. Takao T, Tracey DE, Mitchell M et al. Interleukin-1 receptors in mouse brain: Characterization and neuronal localization. Endocrin 1990; 127:3070-3078.

76. Wong M-L, Licinio J. Localization of interleukin 1 type I receptor mRNA in rat brain. Neuroimmunomodulation 1994; 1:110-115.

77. Takao T, Nakata H, Tojo C et al. Regulation of interleukin-1 receptors and hypothalamic-pituitary-adrenal axis by lipopolysaccharide treatment in the mouse. Brain Res 1994; 649:265-270.

78. Licinio J, Wong ML, Gold PN. Localizatioin of interleukin-1 receptor antagonist mRNA in rat brain. Endocrinol 1991; 129:562-564.

79. Bartfai T, Schultzberg M. Cytokines in neuronal cell types. Neurochem Int 1993; 22:435-444.

80. Benveniste EN. Inflammatory cytokines within the central nervous system: Sources, function, and mechanism of action. Cell Physiol 1992; 32:C1-C16.

81. Fontana A, Kristensen F, Dubs R et al. Production of prostaglandin E and an interleukin-1 like factor by cultured astrocytes and C6 glioma cells. J Immunol 1982; 129:2413-2419.

82. Dinarello CA. Interleukin-1. Rev Infect Dis 1984; 6:51-95.

83. Dinarello CA. Interleukin-1 and interleukin-1 antagonism. Blood 1991; 77:1627-1652.

84. Le J, Vilcek J. Biology of disease. Tumor necrosis factor and interleukin-1: Cytokines with multiple overlapping biological activities. Lab Invest 1987; 56:234-247.

85. Saperas E, Yang H, Tache Y. Interleukin-1β acts at hypothalamic sites to inhibit gastric acid secretion in rats. Am J Physiol 1992; 263:G414-G418.

86. Moldofsky H, Lue FA, Eisen J et al. The relationship of interleukin-1 and immune functions to sleep in humans. Psychosom Med 1986; 48:309-318.

87. Gudewill S, Pollmacher T, Vedder H et al. Nocturnal plasma levels of cytokines in healthy men. Eur Arch Psychiatry Clin Neurosci 1992; 242:53-56.

88. Uthgenannt D, Schoolmann D, Pietrosky R et al. Effects of sleep on the production of cytokines in humans. Psychosomotic Med 1995; 57:97-104.

89. Moldofsky H, Lue F, Davidson JR et al. Effects of sleep deprivation on human immune functions. FASEB J 1989; 3:1972-1977.

90. Banks WA, Ortiz L, Plotkin SR et al. Human interleukin (IL-) 1α, murine IL-1α and murine IL-1β are transported from blood to brain in the mouse by a shared saturable mechanism. J Pharmacol Exp Ther 1991; 259:988-996

91. Blatteis CM. Role of the OVLT in the febrile response to circulating pyrogens. In: Ermisch A, Landraf R, Rahle HJ, eds. Progress in Brain Res. Amsterdam: Elsevier, 1992; 91:409-412.

92. Watkins LR, Goehler LE, Relton JK et al. Blockade of interleukin-1 induced hyperthemia by subdiaphargmatic vasotomy: evidence for vagal mediation of immune-brain communication. Neurosci Lett 1995; 183:27-31.

93. Lue FA, Bail M, Jephthah-Ocholo J et al. Sleep and cerebrospinal fluid interleukin-1 like activity in the cat. Intern J Neurosci 1988; 42:179-183.

94. Fang J, Renegar KB, Kapás L et al. The IL-1 Type I receptor and the TNF 55 kDa receptor are involved in sleep regulation. J Sleep Res 1996; 5:62.

95. Krueger JM, Walter J, Dinarello CA et al. Sleep-promoting effects of endogenous pyrogen (interleukin-1). Am J Physiol 1984; 246:R994-R999.

96. Shoham S, Davenne D, Cady AB et al. Recombinant tumor necrosis factor and interleukin 1 enhance slow-wave sleep. Am J Physiol 1987; 253:R142-R149.

97. Opp MR, Obál F Jr, Krueger JM. Interleukin-1 alters rat sleep: Temporal and dose-related effects. Am J Physiol 1991; 260:R52-R58.

98. Tobler I, Borbély AA, Schwyzer M et al. Interleukin-1 derived from astrocytes enhances slow-wave activity in sleep EEG of the rat. Eur J Pharmacol 1984; 104:191-192.

99. Susic V, Totic S. "Recovery" function of sleep: Effects of purified human interleukin-1 on the sleep and febrile response of cats. Met Brain Dis 1989; 4:73-80.

100. Friedman EM, Boinski S, Coe CL. Interleukin-1 induces sleep-like behavior and alters cell structure in juvenile rhesus macaques. Am J Primatol

101. Obál F Jr, Opp M, Cady AB et al. Interleukin 1α and an interleukin 1β fragment are somnogenic. Am J Physiol 1990; 259:R439-R446.

102. Opp MR, Postlethwaite AE, Seyer JM et al. Interleukin 1 receptor antagonist blocks somnogenic and pyrogenic responses to an interleukin-1 fragment. Proc Natl Acad Sci USA 1992; 89:3726-3730.

103. Hong L, Imeri L, Opp MR et al. Intercellular adhesion molecule-1 expression induced by IL-1 beta or an IL-1 beta fragment is blocked by an IL-1 receptor antagonist and a soluble IL-1 receptor, J Neuroimmunol 1993; 44:163-170.

104. Grenfell S, Smithers N, Witham S et al. Analysis of mutations in the putative nuclear localization sequence of interleukin-1β, Biochem J 1991; 280:111-116.

105. Opp MR, Krueger JM. Anti-interleukin-1β reduces sleep and sleep rebound after sleep deprivation in rats. Am J Physiol 1994; 266:R688-R695.

106. Opp MR, Krueger JM. Interleukin-1 is involved in responses to sleep deprivation in the rabbit. Brain Res 1994; 639:57-65.

107. Opp MR, Krueger JM. Interleukin 1 receptor antagonist blocks interleukin 1-induced sleep and fever. Am J Physiol 1991; 260:R453-R457.

108. Takahashi S, Kapás L, Fang J et al. An interleukin-1 receptor fragment inhibits spontaneous sleep and muramyl dipeptide-induced sleep in rabbits. Am J Physiol 1996;271:R101-R108.

109. Takahashi S, Kapás L, Hansen M et al. An interleukin-1 (IL-1) soluble receptor fragment inhibits IL-1β-induced sleep and non-rapid-eye-movement sleep rebound after sleep deprivation in rabbits. Sleep Res 1995; 24A:457.

110. Imeri L, Opp MR, Krueger JM. An IL-1 receptor and an IL-1 receptor antagonist attenuate muramyl dipeptide- and IL-1-induced sleep and fever. Am J Physiol 1993; 265:R907-R913.

111. Opp MR, Obál F Jr, Krueger JM. Effects of α-MSH on sleep, behavior, and brain temperature: Interactions with IL-1. Am J Physiol 1988; 255:R914-R922.

112. Krueger JM, Kapás L, Opp MR et al. Prostaglandins E_2 and D_2 have little effect on rabbit sleep. Physiol and Behav 1992; 51:481-485.

113. Opp M, Obál F Jr, Krueger JM. Corticotropin-releasing factor attenuates interleukin 1-induced sleep and fever in rabbits. Am J Physiol 1989; 257:R528-R535.

114. Opp MR, Smith EM, Hughes TK. Interleukin-10 (cytokine synthesis inhibitor factor) acts in the central nervous system of rats to reduce sleep. J Neuroimmunol 1995; 60:165-168.

115. Breder CD, Tsujimoto M, Terano Y et al. Distribution and characterization of tumor necrosis factor-α-like immunoreactivity in the murine central nervous system. J Comp Neurology 1993; 337:543-567.

116. Hunt JS, Chen HL, Hu XL et al. Tumor necrosis factor-alpha gene expression in the tissues of normal mice. Cytokine 1992; 4:340-346.

117. Breder CD, Hazuka C, Ghayur T et al. Regional induction of tumor necrosis factor α expression in the mouse brain after systemic lipopolysac-

charide administration. Proc Natl Acad Sci USA 1994; 91:11393-11397.

118. Lieberman AP, Pitha PM, Shin HS et al. Production of tumor necrosis factor and other cytokines by astrocytes stimulated with lipopolysaccharide or neurotropic virus. Proc Natl Acad Sci USA 1989; 399:608-612.

119. Flegenhauer K. A soluble form of tumor necrosis factor receptor in cerebrospinal fluid and serum of HIV-1 associated myclopthy and other neurological disease. J Neurology 1995; 242:239-242.

120. Burns TM, Clough JA, Klein RM et al. Developmental regulation of cytokine expression in the mouse brain. Growth Factors 1993, 9:253-258.

121. Gendron RL, Nestel FP, Lapp WS et al. Expression of tumor necrosis factor alpha in the developing nervous system. Int J Neurosci 1991; 60:129-136.

122. Merrill JE. Tumor necrosis factor alpha, interleukin-1 and related cytokines in brain development: Normal and pathological. Dev Neurosci 1992; 14:1-10.

123. Grassi F, Mileo AM, Monaco L et al. TNF-α increases the frequency of spontaneous miniature synaptic currents in cultured rat hippocampal neurons. Brain Res 1994; 659:226-230.

124. Plata-Salaman CR, Oomura Y, Kai Y. Tumor necrosis factor and interleukin-1β: Suppression of food intake by direct action in the central nervous system. Brain Res 1988; 106-114.

125. Shibata M, Blatteis CM. Human recombinant tumor necrosis factor and interferon affect the activity of neurons in the organum vasculosum laminae terminalis. Brain Res 1991: 562:323-326.

126. Monti JM. Involvement of histamine in the control of the waking state. Life Sci 1993; 53:1331-1338.

127. Alvarez XA, Franco A, Fernandez-Novoa L et al. Effects of neurotoxic lesion in histaminergic neurons on brain tumor necrosis factor levels. Agents Actions 1994; 41:C70-C72.

128. Shoham S, Blatteis CM, Krueger JM. Effects of preoptic area lesions on muramyl dipeptide-induced sleep and fever. Brain Res 1989; 476:396-399.

129. Leeper-Woodford SK, Fisher BJ, Sugerman HJ et al. Pharmacologic reduction in tumor necrosis factor activity of pulmonary alveolar macrophages. Am J Respir Cell Mol Biol 1993; 8:169-175.

130. Takahashi S, Tooley D, Kapás L et al. Inhibition of tumor necrosis factor suppresses sleep in rabbits. Pflügers Arch 1995;431:155-160.

131. Takahashi S, Kapás L, Seyer JM et al. Inhibition of tumor necrosis factor attenuates physiological sleep in rabbits. Neuro Report 1996; 7:642-646.

132. Nistico G, De Sarro G, Rotiroti D. Behavioural and electrocortical spectrum power changes of interleukins and tumor necrosis factor after their microinfusion into different areas of the brain. In: Smirne S et al. Sleep Hormones and Immunological System. Masson: Milano, 1992:11-22.

133. Kapás L, Krueger JM. Tumor necrosis factor-β induces sleep, fever, and anorexia. Am J Physiol 1992; 263:R703-R707.

134. Kapás L, Hong L, Cady AB et al. Somnogenic, pyrogenic, and anorectic activities of tumor necrosis factor-α and TNF-α fragments. Am J Physiol

1992; 263:R708-R715.

135. Takahashi S, Kapás L, Fang J et al. Anti-tumor necrosis factor antibody suppresses sleep in rats and rabbits. Brain Res 1995; 690: 241-244.

136. Takahashi S, Hansen DM, Seyer JM et al. A soluble tumor necrosis factor (TNF) receptor fragment, TNF-R-(159-178), suppresses normal sleep and muramyl dipeptide (MDP)-induced sleep in rabbits. Sleep Res 1995; 24:18.

137. Takahashi S, Kapás L, Seyer JM et al. Inhibition of tumor necrosis factor attenuates the sleep responses induced by increases in ambient temperature in rabbits. Sleep Res 1996;25:33.

138. Darko DF, Mitler MM, Henriksen SS Lentiviral infection immune response peptides and sleep. Adv in Neuroimmunol 1995; 5:57-77.

139. Yamasu K, Shimada Y, Sakaizumi M et al. Activation of the systemic production of tumor necrosis factor after exposure to acute stress. Eur Cytokine Netwk 1992; 3:391-398.

140. Hohagen F, Timmer J, Weyerbrock A et al. Cytokine production during sleep and wakefulness and its relationship to cortisol in healthy humans. Neuropsychobiology 1993; 28:9-16.

141. Bachwich PR, Chensue SW, Larrick JW et al. Tumor necrosis factor stimulates interleukin-1 and prostaglandin E_2 production in resting macrophages. Biochem Biophys Res Comm 1986; 136:94-101.

142. Takahashi S, Kapás L, Fang J et al. Somnogenic relationships between interleukin-1 and tumor necrosis factor. Sleep Res 1996;25:31.

143. Zhang J, Cousens LS, Barr PJ et al. Three-dimensional structure of human basic fibroblast growth factor, a structural homolog of interleukin-1β. Proc Natl Acad Sci USA 1991; 88:3446-3450.

144. Eriksson AE, Cousens LS, Weaver LH et al. Three-dimensional structure of human basic fibroblast growth factor. Proc Natl Acad Sci USA 1991; 88:3441-3445.

145. Zhu X, Komiya H, Chirino A et al. Three-dimensional structures of acidic and basic fibroblast growth factors. Science 1991; 251:90-95.

146. Zeytin FN, Rusk SF, De Lellis R. Growth hormone-releasing factor and fibroblast growth factor regulate somatostatin gene expression. Endocrinology 1988;122: 1133-1136.

147. Oomura Y. Chemical and neuronal control of feeding motivation. Physiol Behav 1988; 44:555-560.

148. Knefati M, Somogyi C, Kapás L et al. Acidic fibroblast growth factor (FGF) but not basic FGF induces sleep and fever in rabbits. Am J Physiol 1995; 269:R87-R91.

149. Hilaire ZD, Nicolaidis S. Enhancement of slow-wave sleep parallel to the satiating effect of acidic fibroblast growth factor in rats. Brain Res Bull 1992; 29:525-528.

150. Sweetman PM, Sanon HR, White LA et al. Differential effects of acidic and basic fibroblast growth factor on spinal cord cholinergic, GABAergic and glutamatergic neurons. J Neurochemistry 1991; 57:237-249.

151. Engel J, Bohn MC. The neurotrophic effects of fibroblast growth factors

on dopaminergic neurons in vitro are mediated by mesencephalic glia. J Neurosci 1991; 11(10):3070-3078.

152. Yoshida K, Gage FH. Fibroblast growth factors stimulate nerve growth factor synthesis and secretion by astrocytes. Brain Res 1991; 538:118-126.

153. De Maeyer E, De Maeyer-Guignard JC. Interferons and Other Regulatory Cytokines. New York: John Wiley & Sons, 1988.

154. Janicki PK. Binding of human alpha-interferon in the brain tissue membranes of rat. Res Comm Chem Pathol Pharmacol 1992; 75:117-120.

155. Smedley H, Katrak M, Sikora K et al. Neurological effects of recombinant human interferon. Br J Med 1983; 28:262-264.

156. Birmanns B, Saphier D, Abramsky O. α-Interferon modifies cortical EEG activity: dose-dependence and antagonism by naloxone. J Neurol Sci 1990; 100:22-26.

157. De Sarro GB, Masuda Y, Ascioti C et al. Behavioural and ECoG spectrum changes induced by intracerebral infusion of interferons and interleukin 2 in rats are antagonized by naloxone. Neuropharmacol 1990; 29:167-179.

158. Dafny N. Interferon modified EEG and EEG-like activity recorded from sensory, motor, and limbic system structures in freely behaving rats. Neurotoxicology 1983; 4:235-240.

159. Reite M, Landenslager M, Jones J et al. Interferon decreases REMS latency. Biol Psychiat 1987; 22:104-107.

160. Krueger JM, Dinarello CA, Shoham S et al. Interferon alpha-2 enhances slow-wave sleep in rabbits. Int J Immunopharmacol 1987; 9:23-30.

161. Kimura M, Majde JA, Toth LA et al. Somnogenic effects of rabbit and human recombinant interferons in rabbits. Am J Physiol 1994; 267:R53-R61.

162. Majde JA. Microbial cell wall contaminants in peptides: A potential source of physiological artifacts. Peptides 1993; 14:629-632.

163. Thompson JA, Cox WW, Lindgren CG et al. Subcutaneous recombinant gamma interferon in cancer patients: Toxicity, pharmacokinetics and immunomodulatory effects, Cancer Immunol Immunother 1987; 25:47-53.

164. Gerrard TL, Siegal JP, Dyer DR et al. Differential effects of interferon alpha and interferon gamma on interleukin-1 secretion by monocytes. J Immunol 1986; 138:2535-2540.

165. Eizenberg O, Faber-Elman A, Lotan M et al. Interleukin-2 transcripts in human and rodent brains: possible expression by astrocytes. J Neurochem 1995; 64:1928-1936.

166. Petitto JM, Huang Z. Molecular cloning of a partial cDNA of the interleukin-2 receptor-beta in normal mouse brain: in situ localization in the hippocampus and expression by neuroblastoma cells. Brain Res 1994; 650:140-145.

167. Opp MR, Krueger JM. Somnogenic actions of interleukin-2: Real or artifact. 1994; Sleep Res 23:26.

168. Opp M, Obál F Jr, Cady AB et al. Interleukin-6 is pyrogenic but not somnogenic. Physiol Behav 1989; 45:1069-1072.

169. Kapás L, Shibata M, Kimura M et al. Inhibition of nitric oxide synthesis suppresses sleep in rabbits. Am J Physiol 1994; 266:R151-R157.

170. Shoham S, Krueger JM. Muramyl dipeptide-induced sleep and fever: Effects of ambient temperature and time of injections. Am J Physiol 1988; 255:R157-R165.

171. Cannon JG, Dinarello, CA. Increased plasma interleukin-1 activity in women after ovulation. Science 1985; 227:1242-1247.

172. Cannon JG, Kluger MJ. Endogenous pyrogen activity in human plasma after exercise. Science 1983; 220:617-619.

173. Sassin JF. Sleep-related hormones. In: Drucker-Colin RR, McGaugh J ed(s). Neurobiology of Sleep and Memory. New York: Academic Press, 1977:361-372.

174. Takahashi Y. Growth hormone secretion related to sleep and waking rhythm. In: Drucker-Colin R, Shkurobich M, Sterman MB ed(s). The Functions of Sleep. New York: Academic Press, 1979:113-145.

175. Ehlers CL, Reed TK, Henriksen SJ. Effects of corticotropin-releasing factor and growth hormone-releasing factor on sleep and activity in rats. Neuroendocrinology 1986; 42:467-474.

176. Obál F Jr. Effects of peptides (DSIP, DSIP analogues, VIP, GRF and CCK) on sleep in the rat. Clin Neuropharmacol 1986; (S4) 9:459-461.

177. Nistico G, De Sarro GB, Bagetta G et al. Behavioral and electrocortical spectrum power effects of growth hormone releasing factor in rats. Neuropharmacology 1987; 26:75-78.

178. Obál F Jr, Alföldi P, Cady AB et al. Growth hormone-releasing factor enhances sleep in rats and rabbits. Am J Physiol 1988; 255:R310-R316.

179. Steiger A, Guldner J, Hemmeter U, Rothe B et al. Changes of sleep-EEG and nocturnal hormonal secretion under pulsatile application of GHRH and somatostatin. Neuroendocrinol 1992; 56:566-573.

180. Kerkhofs M, Van Cauter E, Van Onderberghen A et al. Growth hormone releasing hormone (GHRH) has sleep promoting effects in man. Am. J. Physiol 1993; 264:E594-E598.

181. Obál F Jr, Payne L, Opp M et al. Growth hormone-releasing hormone antibodies suppress sleep and prevent enhancement of sleep after sleep deprivation. Am J Physiol 1992; 263:R1078-R1085.

182. Obál F Jr, Payne L, Kapás L et al. Inhibition of growth hormone releasing hormone suppresses both sleep and growth hormone secretion in the rat. Brain Res 1991; 557:149-153.

183. Payne L, Obál F Jr, Opp MR et al. Stimulation and inhibition of growth hormone secretion by interleukin-1β: The involvement of growth hormone-releasing hormone. Neuroendocrinology 1992; 56:118-123.

184. Rettori V, Jurcovicova J, McCann SM. Central action of interleukin-1 in altering the release of TSH, growth hormone and prolactin in the male rat. J Neurosci Res 1987; 18:179-193.

185. Nash AD, Brandon MR, Bello PA. Effects of tumor necrosis factor-alpha on growth hormone and interleukin-6 mRNA in ovine pituitary cells. Mol Cell Endocrinol 1992; 84:31-37.

186. Honnegger J, Spagnoli A, D'urso R et al. Interleukin-1 beta modulates the acute release of growth hormone releasing hormone and somatostatin from rat hypothalamus in vitro whereas tumor necrosis factor and interleukin-6 have no effect. Endocrinol 1991; 129:1275-1282.

187. Payne LC, Obál F Jr, Krueger JM. Hypothalamic releasing hormones mediating the effects of interleukin-1 on sleep. J Cell Biochem 1993; 53:309-313.

188. Obál F Jr, Fang J, Payne LC et al. Growth hormone-releasing hormone (GHRH) mediates the sleep promoting activity of interleukin-1 (IL-1) in rats. Neuroendocrinology 1995; 61:559-565.

189. Akahoshi T, Oppenheim JJ, Matsushima K. Induction of IL-1 receptor expression on fibrocytes by glucocorticoid hormone, prostaglandins and interleukin-1. J Leukocyte Biol 1987; 42:579.

190. Besedovsky H, Del Rey A, Sorkin E et al. Immunoregulatory feedback between interleukin-1 and glucocorticoid hormones. Science 1986; 233:652-654.

191. Obál FJr, Opp M, Sáry G, Krueger JM. Endocrine mechanisms in sleep regulation. In: Inoué S, Krueger JM ed(s). Endogenous Sleep Factors. The Hague: SPB Academic Publishing bv, 1990: 109-120.

192. Sawada M, Hara N, Maeno T. Ionic mechanism of the outward current induced by extracellular ejection of interleukin-1 onto identified neurons of Aplysia. Brain Res 1991; 545:248-256.

193. Sawada M, Hara N, Ichinose M. Interleukin-2 inhibits the GABA-induced Cl⁻ current in identified Aplysia neurons. J Neurosci Res 1992; 33:461-465.

194. Palazzolo DL, Quadri SK. Interleukin-1 inhibits serotonin release from the hypothalamus in vitro. Life Sci 1992; 51:1797-1802.

195. Gemma C, Ghezzi P, de Simoni M.G. Activation of the hypothalamic serotoninergic system by central interleukin-1. Eur J Pharmacol 1991; 209:139-140.

196. Monhankumar PS, Thyagarajan S, Quadri SK Interleukin-1 beta increases 5-hydroxyindoleacetic acid release in the hypothalamus in vivo. Brain Res Bull 1993; 31:745-748.

197. Sawada M, Hara N, Maeno T. Reduction of the acetylcholine-induced K⁺ current in identified Aplysia neurons by human interleukin-1 and interleukin-2. Cell Mol Neurobiol 1992; 12:439-445.

198. Schultzberg M, Andersson C, Unden A et al. Interleukin-1 in adrenal chromaffin cells, Neurosci 1989; 30:805-810.

199. Soliven B, Albert J. Tumor necrosis factor modulates Ca²⁺ currents in cultured sympathetic neurons. J Neurosci 1992; 12:2665-2671.

200. Koike H, Saito H, Matsuki N. Effect of fibroblast growth factors on calcium currents in acutely isolated neuronal cells from rat ventromedial hypothalamus. Neurosci Lett 1993; 150:57-60.

201. Silverman DHS, Iman K, Karnovsky ML. Muramyl peptide/serotonin receptors in brain-derived preparations. Peptide Res 1989; 2:338-344.

202. Miller LG, Fahey JM. Interleukin-1 modulates GABAergic and gluta-

matergic function in brain. Ann N Y Acad Sci 1994; 739:292-298.

203. Schmidt HHHW, Lohmann SM, Walter U. The nitric oxide and cGMP signal transduction system: Regulation and mechanism of action. Biochem Biophys Acta 1993; 1178:153-175.

204. Rettori V, Belova N, Yu WH et al. Role of nitric oxide in control of growth hormone release in the rat. Neuroimmunomodulation 1994; 1:195-200.

205. Kapás L, Fang J, Krueger JM. Inhibition of nitric oxide synthesis inhibits rat sleep. Brain Res 1994; 664:189-196.

206. Dugovic C, Van deBroeck WAE, DeRyck M et al. Inhibition of nitric oxide synthesis induces opposite effects on deep slow wave and paradoxical sleep in rats. Sleep Res 1995; 24A:119.

207. Kapás L, Krueger JM. The effects of a nitric oxide donor, S-nitroso-N-acetylpenicillamine (SNAP) on sleep in rats. Sleep Res 1994; 23:11.

208. Ayers NA, Kapás L, Krueger JM. Circadian variation of nitric oxide synthase activity and cytosolic protein levels in rat brain. Brain Res 1996;707:127-130.

209. Shibata M. Hypothalamic neuronal responses to cytokines. Yale J Biol and Med 1990; 63:147-156.

210. Shibata M, Blatteis CM. Differential effects of cytokines or thermosensitive neurons in guinea pig preoptic area slices. Am J Physiol 1991; 261:R1096-R1103.

211. Plata-Salamán CR. Immunoregulators in the nervous system. Neurosci Biobehav Rev 1991; 15:185-215.

212. Katsuki H, Nakai S, Hirai Y et al. Interleukin-1β inhibits long-term potentiation in the CA3 region of mouse hippocampal slices. Eur J Pharmacol 1990; 181:323-326.

213. Zeise ML, Madambo S, Siggins GR. Interleukin-1β increases synaptic inhibition in rat hippocampal pyramidal neurons in vitro. Regulatory Peptides 1992; 39:1-7.

214. Edelman GM. Neural Darwinism. New York: Basic Books, 1987.

215. Hebb DO. The Organization of Behavior. New York: Wiley, 1949.

216. Steriade M, Curro Dossi R, Nunez A. Network modulation of a slow intrinsic oscillation of cat thalamocortical neurons implicated in sleep delta waves: Cortically induced synchronization and brainstem cholinergic suppression. J Neurosci 1991; 11:3200-3217.

═══ CHAPTER 5 ═══

CYTOKINE ACTIONS ON FEVER

Matthew J. Kluger, Lisa R. Leon, Wieslaw Kozak,
Dariusz Soszynski and Carole A. Conn

FEVER

Contact with pathogens or simply tissue injury (e.g., sterile abscess) causes the release of cytokines from different cells within the body. Proinflammatory cytokines (e.g., interleukin-1 [IL-1] and IL-6) cause the thermoregulatory set-point to rise at the hypothalamus, and body temperature increases as the result of a coordinated series of physiological and behavioral responses. In humans, fever is accompanied by physiological responses (e.g., peripheral vasoconstriction and shivering) and behavioral responses (e.g., curling into a fetal position, wearing additional clothing and soaking in a warm bath).

Although not all fevers have been shown to be beneficial, there is overwhelming evidence that fever evolved as a host defense response, which reduces morbidity and mortality (for example, see refs. 1, 2). This interaction between the central nervous and immune systems is outlined in Figure 5.1.

In this chapter we describe the evidence that specific cytokines, IL-1, IL-6 and macrophage inflammatory protein-1 (MIP-1) play a role in fever. Tumor necrosis factor α (TNF-α) may be both a fever-producer (i.e., an "endogenous pyrogen") and under certain circumstances, an endogenous antipyretic or cryogen that limits the magnitude of fever.

IL-1β AND IL-1α AND FEVER

IL-1 has been considered an endogenous pyrogen ever since it was first discovered approximately 15 years ago. The evidence for IL-1 being involved in fever was initially based on injecting this material into a variety of laboratory animals and demonstrating that this resulted in fever. The problem with this logic is that many substances will cause fever when injected into animals, but not all of these substances are

Cytokines in the Nervous System, edited by Nancy J. Rothwell. © 1996 R.G. Landes Company.

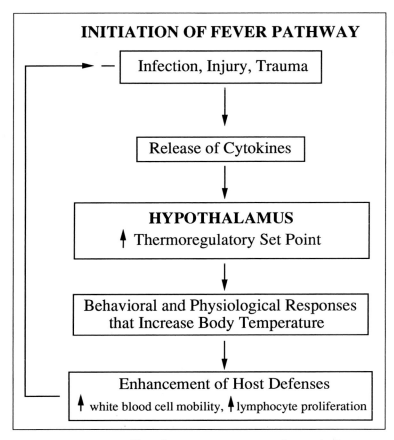

Fig. 5.1. Fever is initiated by infection, injury or trauma (see top). This causes the release of cytokines, which results in a rise in the thermoregulatory set-point in the hypothalamus, a rise in body temperature, and an enhancement of host defenses. This increase in host defenses exerts negative feedback, which reduces the stimuli that initiated febrogenesis (i.e., reduced infection or enhanced tissue repair).

endogenous pyrogens. For example, treatment of patients with high doses of IL-2 often leads to fever and other side-effects.[1] There is, however, little evidence that IL-2 is involved in the pathway between contact with pathogens and the development of fever. Presumably IL-2 exerts its pyrogenicity via intermediary factors such as IL-1, IL-6 or other cytokines. Before a substance can be considered an endogenous mediator of fever, it must fulfill a Koch's postulates-like series of criteria that endocrinologists have used for decades to demonstrate that a putative hormone was indeed responsible for some physiological response.[1] These criteria are particularly important in the area of cytokine biology since so many cytokines trigger the release of other cytokines (the "cytokine cascade"). Injection of what may be a pharmacological dose of IL-1α or some other cytokine may induce fever via induction of other pyrogenic cytokines. The key question is: During infection or inflammation, are IL-1α and IL-1β involved in the fever pathway? The answer, based on using neutralizing antibodies to both of these cytokines, receptor antagonists, soluble receptors, and transgenic mice that lack the gene for IL-1β is no for IL-1α and yes for IL-1β. Below we describe those data supporting this assertion.

EXPERIMENTS USING NEUTRALIZING ANTIBODIES TO IL-1α AND IL-1β

Neutralizing antiserum to IL-1α had no effect on body temperature of lipopolysaccharide (LPS)-injected rats.[3] Long et al[4] and Rothwell et al[5] showed that injection of antiserum to IL-1β reduces fevers in rats caused by injection of LPS. Because of the poor correlation between circulating IL-1 and fever,[1] it is probable that the effects of the neutralizing antibody to IL-1β was exerted either within the central nervous system (CNS), or at some other site outside the circulation (e.g., liver). Since there are dozens of studies showing that IL-1β is produced within the CNS (e.g., ref. 6), we speculated that IL-1β exerts its effects on fever by acting within the CNS. We found that the bioactivity of TNF-α and IL-6 rose in perfusate from the anterior hypothalamus of rats following intraperitoneal (ip) injection of a fever-inducing dose of LPS (50 μg/kg); but there was no increase in hypothalamic concentrations of IL-1.[7] Despite this failure to detect IL-1 in the push-pull perfusate, microinjection of neutralizing antibody to IL-1β into the hypothalamus led to a marked attenuation of fever.[8] These data support the hypothesis that hypothalamic IL-1β (perhaps cell-associated) is responsible for a portion of the rise in body temperature caused by peripheral injection of LPS.

A key question, which we hope to address in future studies, is why we cannot measure IL-1β in the hypothalamic perfusate? As mentioned above, this might be because the cytokine is cell-associated, perhaps membrane-bound or even intracellular. Stitt[9] has argued that IL-1β might act outside the brain proper, perhaps in the organum vasculosum of the lamina terminalis (OVLT), to initiate fever. Our injection of antibody to IL-1β might result in the antibody diffusing the short distance to the OVLT, inhibiting IL-1β in this region, thus blocking the IL-1β-induced rise in hypothalamic IL-6.

EXPERIMENTS USING IL-1 RECEPTOR ANTAGONISTS

Injection ip of IL-1 receptor antagonist (IL-1ra) attenuated LPS-induced fevers in rats.[10] IL-1ra has primary affinity for the IL-1 type I receptor (i.e., it blocks the action of IL-1 by acting on the type I receptor), and Kent et al[11] showed that third ventricular injection of

IL-1ra did not block fevers caused by central injection of IL-1β. These data supported the hypothesis that the pyrogenic effects of IL-1β were not via the type I receptor, and data were presented using the ALVA-42 monoclonal antibody that these fevers might occur via the type II receptor. However, subsequent studies by Gayle et al[12] support the hypothesis that this monoclonal antibody is actually specific for the human major histocompatibility complex class II antigen HLA-DR rather than the IL-1 type II receptor. Sims et al[13] have argued that the type II receptor for IL-1 does not induce intracellular signal and acts as a decoy for IL-1 actions. When we tested the effects of IL-1ra on fever in rats we found results that are opposite to that of Kent et al. Injection of IL-1ra into the lateral ventricle of rats led to attenuation of fever caused by central nervous injection of IL-1β.[14] Injection of the IL-1ra into the lateral ventricle also led to a marked attenuation of fever caused by ip injection of LPS. These data are similar to those recently reported by Dr. Giamal Luheshi, a member of Dr. Nancy Rothwell's laboratory. This attenuation, however, is not total and raises the possibility that an additional IL-1 receptor may exist. At a recent symposium in Japan, Dr. Errol De Souza presented evidence that a third type of IL-1 receptor occurs in the central nervous system, including the anterior hypothalamus. This type III IL-1 receptor may account for a portion of the effects of IL-1 on fever and other acute phase responses.

EXPERIMENTS USING KNOCK-OUT MICE

IL-1β knock-out mice

Injection of LPS (2.5 mg/kg, ip) into IL-1β knock-out mice (mice lacking a functional gene for synthesis of IL-1β) resulted in attenuated fever.[15] Locomotor activity was suppressed in all mice given LPS. With subcutaneous (sc) injection of turpentine, control mice developed large fevers whereas IL-1β knock-out mice developed no fever (Fig. 5.2).[16] Furthermore, whereas the control mice became inactive for 24 hours after injection of turpentine, the knock-out mice showed no deficit in locomotor activity. These data support the hypotheses that induction of a localized inflammation (i.e., sc injection of turpentine) causes fever (and a reduction in activity) exclusively via IL-1β, whereas the induction of systemic inflammation (i.e., ip injection of LPS) exerts its actions only partially via IL-1β.

IL-6 knock-out mice

Injection of LPS (50 μg/kg, ip) and recombinant IL-1β (10 μg/kg ip) failed to produce fever in IL-6 knock-out mice.[17] Since these transgenic mice produce normal elevations of IL-1α in response to LPS,[18] these data support the hypothesis that IL-1α does not mediate LPS-induced fever, and implicate IL-1β via IL-6 in the fever pathway.

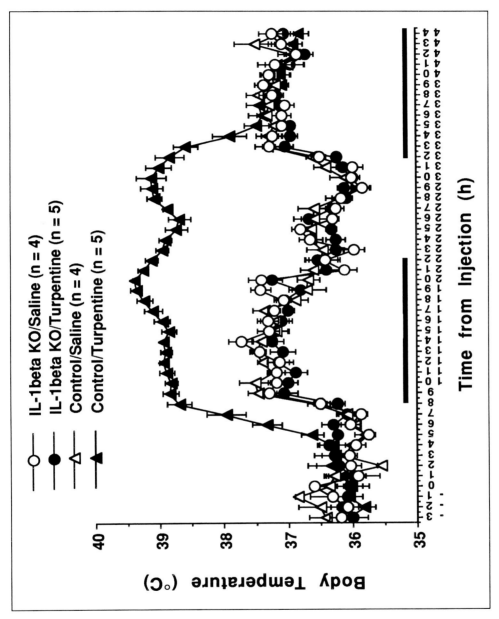

Fig. 5.2. Fever induction in mice following injection of turpentine. Mice were implanted with Mini-Mitter transmitters to monitor body temperature. One group lacked the gene for IL-1β (IL-1β knock-outs); the other group of mice were wild-type controls. They were then injected sc with 100 μl of turpentine or saline. Note that the wild-type controls injected with turpentine developed large fevers whereas the knock-out mice did not develop any fever.[16]

Legend:
○ IL-1beta KO/Saline (n = 4)
● IL-1beta KO/Turpentine (n = 5)
△ Control/Saline (n = 4)
▲ Control/Turpentine (n = 5)

X-axis: Time from Injection (h)
Y-axis: Body Temperature (°C)

IL-1 type I receptor knock-out mice

In ongoing studies, Lisa Leon and others within our group are studying the febrile responses of mice that lack the gene for IL-1 type I receptors (Immunex). These mice develop normal fevers after ip injection of a high (2.5 mg/kg) and low (50 µg/kg) doses of LPS, but fail to develop any fever in response to subcutaneous injection of turpentine. These data indicate that a localized inflammation causes fever exclusively via the IL-1 type I receptor, whereas the induction of systemic inflammation (i.e., ip injection of LPS) does not occur via the IL-1 type I receptor in mice. Based on the attenuated fever to LPS in IL-1β knock-out mice and the normal febrile response to LPS in mice lacking the IL-1 type I receptor, we conclude that IL-1β mediates a portion of LPS fever, but this is probably not via the type I receptor in mice. It is possible that in mice LPS-induced fevers occur largely via the proposed type III receptor. It remains to be tested whether IL-1ra administered into the CNS of rats (or other species) attenuates fevers caused by local inflammation (i.e., sc injection of turpentine).

SUMMARY OF IL-1 DATA

The following conclusions can be drawn from experiments using rats and mice as animal models to dissect the role of IL-1 in fever:

1. IL-1α is not involved in LPS-induced fever.
2. IL-1β is involved in a significant portion of LPS-induced fever. Other mediators must play a role in LPS-induced fever. IL-1β is an essential cytokine for fevers caused by sc injection of turpentine.
3. Based on data generated by microinjection of IL-1ra into the lateral ventricle of rats, IL-1β exerts its pyrogenic activity largely via the IL-1 type I receptor.
4. IL-1 type I receptors are essential for fevers induced by sc injection of turpentine, but not for ip injection of LPS in *mice*. Additional studies are necessary to determine the apparent differences between rats and mice with respect to the roles of the IL-1 type I, IL-1 type III (and possibly other) receptors in LPS-induced fever.

IL-6 AND FEVER

Nijsten et al[19] showed that the body temperature of patients with severe burns correlated with plasma concentration of IL-6, and concluded that "IL-6 plays a causal role in the acute phase response." Helle et al[20] then showed that purified IL-6 induces fever when injected intravenously (iv) in rabbits. However, as in many studies using other cytokines, it was not shown whether the amount required to cause fever (0.5 to 5 µg/kg) achieved plasma concentrations that were within the normal physiologic or pathophysiologic range. LeMay et al[21] presented evidence that IL-6 causes fever when injected icv in rats, but again, there was no indication that the dose of IL-6 necessary to cause fever had any physiological relevance. Similar data were reported by Sakata et al[22] for rabbits. As in the study by Helle et al[20] and in numerous subsequent studies, the dose of IL-6 necessary to induce fever far surpasses that required by IL-1 to initiate a fever. It is important to realize, however, that the circulating or brain concentrations of IL-6 are extremely high, whereas IL-1 is generally low to nondetectable following the initiation of an inflammatory response.[21] Using ELISAs Zheng et al[16] showed that plasma levels of IL-1β in mice are approximately 0.6 ng/ml following injection of LPS, whereas plasma levels of IL-6 were 193 ng/ml.

Experiments using antibodies to IL-6 and IL-1β

Rothwell et al[23] showed in rats that icv injection of an antibody to rat IL-6 inhibited the febrile effects of IL-1β injected icv or LPS injected ip. Frei et al[24] reported that IL-6 is produced in brain tissue during a variety of CNS infections, and these findings have since been confirmed by many other groups. It is possible that it is brain IL-6, rather than plasma IL-6, that is important in the development of fever. We have found that systemic (ip) pretreatment of rats with antiserum to IL-1β not only blocks a significant portion of the fever in response to injected LPS, but the antiserum to IL-1β also significantly attenuates the rise in IL-6 in both plasma and cerebrospinal fluid.[25] These data concur with the results of many in vitro studies indicating that IL-1 can induce the production of IL-6, and support the hypothesis that IL-1β induces fever via the production of IL-6.

Further support for the above hypothesis comes from a study by Klir et al[8] in which antibody to IL-1β was microinjected unilaterally into the anterior hypothalamus. This not only blocked 58% of the rise in body temperature to ip injected LPS, but also completely abrogated the rise in IL-6 (97% inhibition) observed in the push-pull perfusate from the unilateral site of the anterior hypothalamus. The injection of the antibody to IL-1β into the anterior hypothalamus had no effect on circulating IL-6 levels. Thus, we concluded that the IL-6 important in the development of LPS-induced fever was being produced directly within the central nervous system in response to local IL-1β mediation. Klir et al.[7] earlier had shown that infusion of a physiologically-relevant concentration of IL-6 into the anterior hypothalamus of rats caused fevers similar in magnitude to that observed following ip injection of LPS.

Studies using IL-6 knock-out mice

As mentioned above, Chai et al[17] have shown that IL-6 deficient mice (lacking the gene for IL-6) fail to develop fevers in response to ip injection of LPS (50 µg/kg) and to recombinant mouse IL-1β (10 µg/kg). Unpublished results from our laboratory indicate that these knock-out mice develop normal fevers to the ip injection of 2.5 mg/kg LPS. Thus, whereas IL-6 is critically important for fevers to low doses of LPS, high doses of LPS induce fever in mice via an IL-6-independent pathway. Interestingly, as in IL-1β knock-out mice, injection of turpentine in IL-6 knock-out mice does not lead to fever (Kozak et al, unpublished data).

SUMMARY OF IL-6 DATA

1. IL-6 has all the characteristics of an endogenous pyrogen. For example: (a) IL-6 rises in the anterior hypothalamus following ip injection of LPS; (b) infusion of IL-6 into the anterior hypothalamus at a concentration that simulates that seen following ip injection of LPS causes fever; and (c) injection of neutralizing antibody to IL-6 icv attenuates fevers caused by ip injection of LPS.

2. It is probable that IL-1β mediates fever via an increase in intrahypothalamic IL-6.

MIP-1 AND FEVER

Davatelis et al[26] showed that a crude preparation of macrophage inflammatory protein-1 (MIP-1) led to prostaglandin-independent fevers. Miñano et al[27] then showed that MIP-1 led to fevers when injected intrahypothalamically into rats. Pretreatment of the injection site with indomethacin failed to prevent the development of fever. MIP-1 consists of two peptides, each about 8 kDa in size. Myers et al[28] showed that injection of both MIP-1α and MIP-1β caused fevers in rats.

These data indicate that MIP-1 is capable of inducing fever when injected into rats, and that these fevers are not dependent on prostaglandins. But is MIP-1 actually in the fever pathway?

Studies using antibodies to MIP-1

Miñano et al[29] have shown that microinjection of goat anti-mouse MIP-1β antibody into the preoptic anterior hypothalamus of rats markedly suppressed fever in response to an ip injection of LPS. The attenuation of fever occurred when the antibody was injected at the time of injection of LPS or even three hours after the injection of LPS when fever was close to its maximum.

These data lead to a paradox. Antibody to MIP-1β blocks fevers caused by peripheral injection of LPS. It is well known that cyclooxygenase inhibitors block LPS-induced fevers, yet MIP-1 fevers are presumably not mediated by prostaglandins. However, as described earlier in this review, studies using antibodies to IL-1β (or using IL-1β knock-out mice) indicate that this leads to the attenuation of LPS-induced fevers. Furthermore, IL-1β fevers are blocked by cyclooxygenase inhibitors. It is possible that injection of LPS or other inflammatory mediators induce the production of several cytokines necessary for the generation of full-blown fever. Neutralization of, for example, either IL-1β or MIP-1β leads to marked attenuation of fever (even though one presumably works via prostaglandins and the other via a prostaglandin-independent pathway).

TNF AND FEVER/ANTIPYRESIS

Michie et al[30] found that intravenous injection of LPS to human patients resulted in fever and an increase in plasma concentration of TNF. They then infused these patients with TNF at a dose that attempted to simulate the plasma concentrations measured following injection of LPS. Their findings support the hypothesis that TNF is a pyrogen in human subjects. Although their results are impressive, two aspects of their study are difficult to interpret. One is that despite continuous infusion of TNF, the plasma concentration of TNF (as determined by ELISA) fell to non-detectable levels after 12 hours. Perhaps the presence of high levels of TNF induces its rapid clearance? Since there are reports of discrepancies between immunoassays and bioassays for measurement of TNF,[31-33] it is also possible that the measurements of TNF in the study by Michie et al[30] do not accurately reflect the biologically active TNF. A second unresolved aspect of the study by Michie et al is that the patients developed fevers despite pre-treatment with indomethacin, a nonsteroidal antiinflammatory drug that has potent antipyretic properties.

In studies from our laboratory, Klir et al[7] measured the bioactivities of TNF and IL-6 in the anterior hypothalamus (using push-pull perfusion) of rats injected ip with LPS. The hypothalamic bioactivities of TNF and IL-6 rose. When we simulated these hypothalamic concentrations of cytokines by slowly infusing IL-6 or TNF intrahypothalamically, IL-6 induced fever, whereas TNF had no effect on body temperature. Infusion of higher (probably *pharmacological*) doses of TNF did produce fever. (As mentioned earlier, infusion of physiological levels of IL-6 led to fever in these rats).

Endogenous antipyretics

There are also considerable data indicating that a large number of endogenously produced factors attenuate the rise in body temperature. These endogenous antipyretics or cryogens include such hormones as αMSH [34], arginine vasopressin,[35] and glucocorticoids.[36] In addition, while many cytokines are proinflammatory (e.g., IL-1), others have immunosuppressive actions (e.g., IL-10).[37-39] It is possible that some of these may exert antipyretic or cryogenic activity during infection. Although TNF has often been considered an endogenous pyrogen, much of the data reported in the literature support another interpretation, namely that under many circumstances TNF is an endogenous cryogen.

Pentoxifylline, TNF and fever

Zabel et al[40] have shown that the drug pentoxifylline totally abolished the rise in plasma TNF concentration in human volunteers given LPS intravenously without influencing the course of their fever, their "flu-like symptoms" and leukocytosis. We found similar results in our experiments in LPS-injected rats that were pretreated with varying doses of pentoxifylline.[41] At moderate doses, this drug did not influence fever, but did significantly reduce the plasma concentration of TNF in LPS-injected rats. Taken together, these data support the hypothesis that circulating TNF is not essential for the development of LPS-induced fever.

Experiments using antibodies to TNF

Rats were pretreated with antiserum to TNF in amounts that completely blocked any rise in plasma TNF, as determined by the sensitive WEHI assay.[4,42] When these rats injected with antiserum either ip or iv were then injected with LPS ip, the resultant fevers were significantly enhanced, rather than suppressed. Our data are consistent with the hypothesis that in the rat TNF may be an endogenous antipyretic that limits the magnitude of fever. Similar findings were found by Smith and Kluger[43] in rats inoculated with an MCA sarcoma. Antiserum to TNF prevented the tumor-induced fall in body temperature. In mice injected with a large dose of LPS (2.5 mg/kg) both antiserum to TNF and soluble TNF receptor blocked the fall in body temperature that preceded the onset of fever.[44] Derijk and Berkenbosch[45] found a similar attenuation in the fall in body temperature in rats injected with antiserum to TNF followed by injection of a high dose of LPS (500 µg/kg). In a more recent study, Klir et al[46] showed that injection of small, non-pyrogenic, doses of TNF ip into rats blocked LPS-induced fevers, confirming earlier findings by Long et al.[47] Kozak et al[44] reported similar results in mice. Based on injections of TNF intracerebroventricularly and intrahypothalamically,[7] our data support the hypothesis that the antipyretic actions of TNF reside at some location outside the CNS.

Nagai et al[48] and Kawasaki et al[49] found that administration of monoclonal antibody to TNF resulted in a significant attenuation of LPS-fever in the rabbit. In these studies, rabbits that were injected with LPS and antibody to TNF developed similar fevers during the first two hours, but there was a significant reduction in the fever of the animals receiving antibody during hours two to four. No data were presented showing that these monoclonal antibodies neutralized TNF bioactivity. In another study, a single iv injection of a murine monoclonal antibody to TNF led to *smaller* fevers in children receiving otherwise conventional therapy for cerebral malaria.[50] However, no evidence was presented indicating that the TNF was neutralized. In fact it was reported in this study that the circulating levels of TNF were actually significantly *higher* in those patients given the antibody. These investigators concluded that the presence of higher TNF in the circulation meant that less TNF could get to the CNS, and thus the fever was reduced. An alternative explanation, based on the data described above showing that injections of non-pyrogenic doses of TNF results in antipyresis, is that TNF is an endogenous cryogen, acting at a level outside the CNS—the higher circulating levels of TNF led to a greater signal to reduce body temperature.

How can we explain the differences between our experiments in rats and mice and those of Nagai et al and Kawasaki et al in rabbits? As mentioned above, in the studies involving injection of monoclonal antibody to TNF in rabbits no data were presented demonstrating that their antibody neutralized TNF bioactivity in vivo. It is possible that injection of this antibody might have increased levels of TNF (as shown in the human study[50]), producing greater antipyresis—thus, resulting in smaller fevers. Of course, assuming the antibody used in the above studies completely neutralized in vivo TNF bioactivity, an alternate hypothesis is that there is a species difference between the rabbit and rat/mouse in the antipyretic actions of TNF. The lack of a TNF negative feedback on fever (and perhaps other acute phase responses) might account for the well-established hypersensitivity of rabbits and humans to the effects of LPS. An argument against there being a species difference in TNF pyrogenicity/cryogenicity is that Mathison et al[51] showed that pretreating rabbits with polyclonal antiserum to TNF resulted in in vivo neutralization of TNF and led to significant protection against LPS-induced hypotension, fibrin deposition, and lethality. Despite this demonstration of the efficacy of their antiserum, fever was not attenuated and actually appeared to be significantly enhanced.

The mechanism of this TNF antipyresis is unknown. Our data support the hypothesis that the antipyretic actions of TNF reside outside the CNS. TNF might induce the production and release of some antipyretic cytokine or other substance from sites outside the CNS, which then crosses from the circulation to the CNS. As mentioned above, we (and others) have not been able to demonstrate any direct CNS fever-modulation when small physiologically relevant doses of TNF, or antibody to TNF, are injected intracerebroventricularly or directly into the anterior hypothalamus. In a recent study, Ebisui et al[52] showed that intravenous injection of antiserum to TNF markedly attenuated the rise in plasma corticosterone induced by injection of LPS in the rat. Since LPS-induced rise in corticosterone results in attenuation of fever,[36] it is possible that TNF's antipyretic action occurs through the release of glucocorticoids. Recent evidence points towards the anterior hypothalamus as the site of glucocorticoid suppression of fever.[53]

Work from Rothwell's group indicates that fevers induced by local inflammation (i.e., turpentine abscess) may actually be mediated by TNF. In unpublished data, rats treated with neutralizing antibody to

TNF developed smaller fevers in response to turpentine abscesses. Thus, depending on the insult, TNF may be either an endogenous cryogen or endogenous pyrogen.

Experiments using TNF soluble receptor

Klir et al[46] found that ip injection of the TNF soluble receptor (TNF:Fc, Immunex) into rats, at a dose that prevents the biological activity of TNF, led to larger LPS-induced fevers. These results were later confirmed in a mouse model of fever by Kozak et al.[44]

Studies using transgenic mice lacking TNF receptors

In ongoing studies in our laboratory, Lisa Leon has examined the role of endogenous TNF receptors in acute phase responses to systemic (LPS) and local (turpentine) inflammatory stimuli.[54] The peripheral injection of a high dose of LPS (2.5 mg/kg, ip) into transgenic mice lacking the p55 (type I) and p75 (type II) TNF receptors (TNFR knock-out mice) resulted in an exacerbation of the early phase of fever (2-15 h) compared to LPS-injected TNFR wild-type mice. The late phase of fever (15-24 h) as well as the lethargy, decrease in body weight and anorexia in response to this dose of LPS were virtually identical in both groups of mice. These results suggest that endogenous TNF-α is involved in the modulation of fever to a high dose of LPS in mice as an endogenous cryogen or antipyretic. These results are similar to those reported by Kozak et al[44] demonstrating the ability of antiserum against TNF-α or soluble TNF receptor (sTNFR) to abolish the early hypothermic phase of fever in Swiss-Webster mice to the same dose of LPS resulting in a more rapid development of fever. As in studies using the transgenic mice deficient in TNF receptors, antiserum to TNF-α and sTNFR did not affect the peak febrile temperature to this high dose of LPS.

In contrast to the high dose of LPS, the peripheral injection of a low dose of LPS (50 µg/kg, ip) into TNFR knock-out and wild-type mice resulted in an approximately 7 h fever that did not differ during any phase of its development between genotypes. Again, lethargy, body weight and food intake were similarly depressed in response to this dose of LPS in both knock-out and wild-type mice. These results suggest that endogenous TNF receptors are differentially involved in the modulation of fever in mice to a peripheral injection of LPS and this modulatory role is dependent upon the dose of LPS injected—the higher the dose of LPS, the more cryogenic is the activity of TNF.

To examine the role of endogenous TNF receptors in the development of acute phase responses to a *local* inflammatory stimulus, TNFR knock-out and wild-type mice were peripherally injected into the left hind limb with turpentine (100 µl/mouse, sc). Our data suggest that endogenous TNF receptors are not involved in acute phase responses to tissue injury as indicated by identical fever, cachexia, and anorectic responses of TNFR wild-type and knock-out mice.[54] These data are in contrast to those reported by Cooper et al[55] demonstrating an attenuating effect of TNF antiserum on fevers and hypermetabolic responses to turpentine in rats. However, some investigators have been unable to detect TNF-α in the circulation during turpentine abscess. Although it has been postulated that the efficacy of TNF-α antiserum treatment on attenuation of turpentine-induced fevers could be due to a compartmentalized action of TNF-α at the local site of inflammation,[56] our data using TNF double receptor knock-out mice do not support this hypothesis since these mice lack both receptors in all tissues of the body. TNF-α antibody treatment in rats[57] and rabbits[58] has also resulted in an attenuation of anorexia in response to turpentine, while the TNFR knock-out and wild-type mice used in our study showed similar, but not significant, reductions in food intake in response to turpentine. Perhaps the absence of TNF receptors in the knock-out mice results in the sustained, exacerbated release of another cytokine, such as IL-1, in response to injection of turpentine that affects fever, body weight and food intake parameters differently than short-term administration of neutralizing agents of TNF-α.

SUMMARY OF TNF DATA

1. TNF can be pyrogenic when infused/injected at high doses into experimental animals or people. Is this the result of the physiological or pharmacological actions of TNF?
2. Treatment of animals with *neutralizing* antibodies to TNF or TNF soluble receptors results in larger fevers, rather than smaller fevers in response to injection of LPS.
3. Treatment of people and rabbits with *non-neutralizing* antibody to TNF, which results in higher circulating levels of TNF, results in attenuated fever. This may be due to the higher concentration of TNF in the circulation.
4. Injection of a low dose of TNF into rats, which does not influence body temperature by itself, attenuates fevers to injection of LPS.
5. Local inflammation induced by turpentine may induce fever via TNF in rats, whereas in mice it appears that TNF plays little or no role in turpentine-induced fevers.

The preponderance of evidence support the hypothesis that endogenously produced TNF is an endogenous antipyretic or cryogen. However, under certain conditions TNF may be an endogenous pyrogen. Much work is still needed to dissect the precise role of TNF in fevers induced by various inflammatory and infectious agents.

MUST CYTOKINES EXERT THEIR PYROGENIC/CRYOGENIC ACTIVITIES AT THE LEVEL OF THE CNS?

Most infections are localized (e.g., peritonitis, gastritis) and occur outside the central nervous system. What is the means by which information regarding status of health is transmitted to the brain? The classical explanation is that the message is transmitted humorally. For example, a localized abscess induces the production of a circulating endogenous pyrogen, which crosses the blood brain barrier at the level of the anterior hypothalamus to induce fever. However, as mentioned earlier in this review, there is little evidence that IL-1 is a circulating cytokine, and fever caused by IL-6 appears to be due to a localized increase in IL-6 in the hypothalamus rather than the result of the elevation in circulating IL-6. Furthermore, the antipyretic action of TNF appears to reside outside the CNS.

Within the past couple of years two groups have shown that signals that influence sickness behavior in rats can be transmitted via the vagus nerve. Bluthé et al[59] demonstrated that ip injection of LPS failed to suppress investigatory behavior in rats that were sub-diaphragmatically vagotomized. Sham-vagotomized rats showed the normal reduction in exploratory behavior following an injection of LPS. Goehler et al[60] generated similar data on conditioned taste aversion. They found that ip administration of either IL-1β or TNF led to avoidance of novel tastes with which they were paired. However, sub-diaphragmatic vagotomy led to an attenuation in this aversion. In a subsequent study, Watkins et al[61] showed that sub-diaphragmatic vagotomy blocked fevers caused by ip injection of IL-1β. Surprisingly, hepatic vagotomy did not lead to any attenuation in fevers caused by injection of IL-1β, suggesting that the pyrogenic effects of IL-1β were not being mediated via afferents originating from the liver. As mentioned in this review, IL-1β is not generally found in the circulation, and it would be informative to determine the effects of vagotomy on fevers caused by LPS or some other inflammatory stimuli (e.g., real infection or turpentine-induced inflammation). Our own data show that administration of antibody to IL-1β within the anterior hypothalamus results in the attenuation of fever to ip injection of LPS. It is possible that an ip injection of LPS stimulates vagal afferents outside the CNS, which raises the concentration of IL-1β within the anterior hypothalamus. In a follow-up study from Dantzer's group, Layé et al[62] showed that sub-diaphragmatic vagotomy blocks the induction of IL-1β mRNA in the mouse brain in response to peripheral injection of LPS. If cytokines or other inflammatory stimuli are triggering fever, in part, via neural pathways originating outside the CNS, then it is possible that endogenous antipyretics (e.g., TNF) might exert their antipyretic actions by inhibition of vagal afferents (or other neural pathways), thus blocking the pyrogenic activity of cytokines such as IL-1β.

SUMMARY

Based on data published within the past few years we believe the model of fever shown in Figure 5.3 may describe fevers induced by bacterial products. In response to an injection of LPS (and possibly during systemic bacterial infections), organs within the abdomen (e.g.,

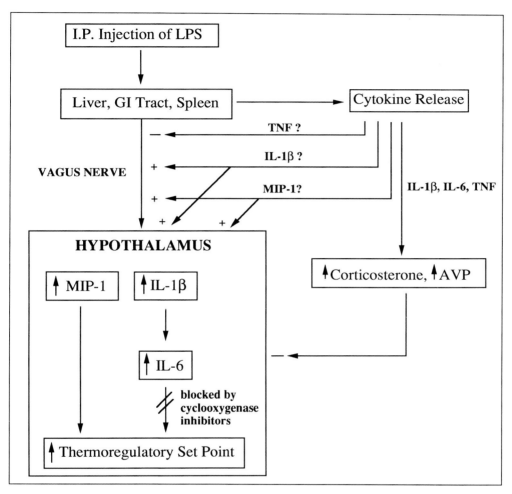

Fig. 5.3. Model of fever based on recent data. Intraperitoneal injection of LPS (top left) induces the production of TNF, IL-6, MIP-1 and other cytokines from various organs in the viscera. These cytokines probably do not directly initiate fever. Some signal (perhaps IL-1β) initiates afferent information traveling from the vagus nerve to the hypothalamus to increase IL-1β, which in turn causes an elevation in hypothalamic IL-6, thus raising the thermoregulatory set-point. Circulating cytokines might modulate fever as follows:
(a) IL-1β (IL-6) may be exerting a stimulatory effect on the vagal afferent information,
(b) TNF may be exerting its antipyretic activity via inhibition of the vagal afferent information to the anterior hypothalamus,
(c) IL-1β, IL-6 and TNFα (both circulating and within the CNS) induce the release of the antipyretic hormones corticosterone[36] and arginine vasopressin,[35]
(d) MIP-1 may be acting directly at the anterior hypothalamus via a non-prostaglandin pathway.

liver, spleen, gastrointestinal tract) are activated to produce cytokines. We know, for example, that the liver is capable of producing and releasing large amounts of IL-6 and TNF into the circulation (for example, in ref. 63). These circulating cytokines are probably not the direct cause of the increase in the thermoregulatory set-point. Based on data presented above, it appears that some of the information required to initiate fever is carried via vagal afferents to the hypothalamus, where it induces a local rise in hypothalamic IL-1β. As a result of our push-pull perfusion studies, we believe that this IL-1β remains cell-associated, and induces a rise in IL-6 within the anterior hypothalamus. It is the rise in IL-6 within this region of the brain that results in the increase in thermoregulatory set-point, leading to the chills and peripheral vasoconstriction associated with the rising phase of fever. In addition, the release of cytokines also results in the secretion of hormones (e.g., corticosterone) that act within the CNS to attenuate fever. Whether this model of fever will prove to be correct for fevers caused by other stimuli (e.g., real bacterial or viral infections) remains to be determined.

REFERENCES

1. Kluger MJ. Fever: role of endogenous pyrogens and cryogens. Physiological Reviews 1991; 71:93-127.
2. Roberts NJ Jr. The immunological consequences of fever, In: Mackowiak PA, ed.: Fever, Basic Mechanisms and Management. Raven Press, 1991:125-142.
3. Long NC, Kluger MJ, Vander AJ. Antiserum against mouse IL-1 alpha does not block stress hyperthermia or LPS fever in the rat. In: Thermoregulation: Research and Clinical Applications. P Lomax, E Schonbaum, eds. 7th International Symposium Pharmacology of Thermoregulation, Odense, Denmark, Karger, Basel. 1988,78-84.
4. Long NC, Otterness I, Kunkel SL, Vander AJ, Kluger MJ. The roles of interleukin-1β and tumor necrosis factor in lipopolysaccharide-fever in the rat. Am J Physiol 1990; 259:R724-R728.
5. Rothwell NJ, Busbridge NJ, Humphray H, Hissey P. Central actions of interleukin-1β on fever and thermogenesis. In: Dinarello CA, Kluger MJ, Powanda MC, Oppenheim JJ (eds): The Physiological and Pathological Effects of Cytokines, Wiley-Liss, Inc., 1990:307-311.
6. De Souza EB. Neurobiology of Cytokines, Parts A and B. Methods in Neurosciences. Academic Press, San Diego, 1993.
7. Klir JJ, Roth J, Szelenyi Z, McClellan JL, Kluger MJ. Role of hypothalamic interleukin-6 and tumor necrosis factor α in LPS-fever in rat. Am J Physiol 1993; 265:R512-R517.
8. Klir JJ, McClellan JL, Kluger MJ. Interleukin-1β causes the increase in anterior hypothalamic interleukin-6 during LPS-induced fever in rats. Am J Physiol 1994; 266:R1845-R1848.
9. Stitt JT. Central regulation of body temperature. In: Exercise, Heat and

Thermoregulation, Perspectives. In: Gisolfi CV, Lamb DR, Nadel ER, eds. Exercise Science and Sports Medicine. Volume 6. Brown and Benchmark, 1993; 1-39.

10. Smith BK, Kluger MJ. Human IL-1 receptor antagonist partially suppresses LPS fever but not plasma levels of IL-6 in the Fischer rat. Am J Physiol 1992; 263:R653-R655.

11. Kent S, Bluthé RM, Dantzer R, Hardwick AJ, Kelley KW, Rothwell NJ, Vannice JL. Different receptor mechanisms mediate the pyrogenic and behavioral effects of interleukin 1. Proc Natl Acad Sci USA 1992; 89:9117-9120.

12. Gayle MA, Sims JE, Dower SK, Slack JL. Monoclonal antibody 1994-01 (also known as ALVA 42) reported to recognize type II IL-1 receptor is specific for HLA-DR alpha and beta chains. Cytokine 1994; 6:83-86.

13. Sims JE, Gayle MA, Slack JL, Alderson MR, Bird TA, Giri JG, Colotta F, Re F, Mantovani A, Shanebeck K, Grabstein KH, Dower SK. Interleukin-1 signaling occurs exclusively via the type I receptor. Proc Natl Acad Sci USA 1993,90:6155-6159.

14. Leon LR, Sims J, Kluger MJ. Attenuation of central interleukin-1β fever following type I but not type II receptor antagonism in rats. FASEB J 1995; 9:A645.

15. Kozak W, Zheng H, Conn CA, Soszynski D, Van der Ploeg L, Kluger MJ. Thermal and behavioral effects of lipopolysaccharide and influenza in interleukin-1β deficient mice. Am J Physiol 1995; 269:R969-R977.

16. Zheng H, Fletcher D, Kozak W, Jiang M, Hofmann K, Conn CA, Soszynski D, Grabiec C, Trumbauer ME, Shaw A, Kostura MJ, Stevens K, Rosen H, North RJ, Chen HY, Tocci MJ, Kluger MJ, Van der Ploeg LHT. Resistance to fever induction and impaired acute-phase response in interleukin-1β deficient mice. Immunity 1995; 3:9-19.

17. Chai Z, Gatti S, Poli V, Bartfai T. IL-6 gene expression is necessary for fever induction in response to LPS in mice: a study on IL-6 deficient mice. In Silvia Gatti, Ph.D. Thesis. Interleukin 1, tumor necrosis factor α and interleukin 6 in the rat central nervous system: production and effects with particular attention to the fever response, Stockholm University. 1994.

18. Fattori E, Capplelletti M, Costa P, Sellitto C, Cantoni L, Carelli M, Faggioni F, Fantuzzi G, Ghezzi P, Poli V. Defective inflammatory response in interleukin 6-deficient mice. J Exp Med 1994; 180:1243-1250.

19. Nijsten MWN, De Groot ER, Ten Duis HJ, Klasen HJ, Hack CE, Aarden LA. Serum levels of interleukin-6 and acute phase response. The Lancet 1987, October 17, 2(8564):921.

20. Helle M, Brakenfood JPJ, De Groot ER, Aarden LA. Interleukin 6 is involved in interleukin 1-induced activities. Eur J Immunol 1988; 18:957-59.

21. LeMay L, Vander AJ, Kluger MJ. The role of IL-6 in fever in the rat. Am J Physiol 1990; 258:R798-R803.

22. Sakata Y, Morimoto A, Long NC, Murakami N. Fever and acute-phase

response induced in rabbits by intravenous and intracerebroventricular injection of interleukin-6. Cytokine 1991; 3:199-203.

23. Rothwell NJ, Busbridge NJ, Lefeuvre RA, Hardwick AJ, Gauldie J, Hopkins SJ. Interleukin-6 is a centrally acting endogenous pyrogen in the rat. Can J Physiol Pharmacol 1991; 69:1465-1469.

24. Frei K, Leist TP, Meager A, Gallo P, Leppert D, Zinkernagel RM, Fontana A. Production of B cell stimulatory factor-2 and interferon in the central nervous system during viral meningitis and encephalitis. J Exp Med 1988; 168:449-453.

25. LeMay L, Otterness I, Vander AJ, Kluger MJ. In vivo evidence that the rise in plasma IL-6 following injection of fever-inducing dose of LPS is mediated by IL-1β. Cytokine 1990; 2:199-204.

26. Davatelis G, Wolpe SD, Sherry B, Dayer J-M, Chicheportiche R, Cerami A. Macrophage inflammatory protein-1: a prostaglandin-independent endogenous pyrogen. Science 1989; 243:1066-1068.

27. Miñano FJ, Sancibrian M, Vizcaino M, Paez X, Davatelis G, Fahey T, Sherry B, Cerami A, and Myers RD. Macrophage inflammatory protein-1: unique action on the hypothalamus to evoke fever. Brain Research Bulletin 1990; 24:849-852.

28. Myers RD, Paex X, Roscoe AK, Sherry B, and Cerami A. Fever and feeding: differential actions of macrophage inflammatory protein-1 (MIP-1), MIP-1α and MIP-1β on rat hypothalamus. Neurochemical Research 1993; 18:667-673.

29. Miñano FJ, Alonso-Fernández A, Benamar K, Myers RD, Sancibrián M, Ruiz RM, and Armengol JA. Macrophage inflammatory protein-1β (MIP-1β) produced endogenously in brain during *E. coli* fever in rats. Eur J Neuroscience 1996;8:424-428.

30. Michie HR, Spriggs DR, Manogue KR, Sherman ML, Revhaug A, O'Dwyer ST, Arthur K, Dinarello CA, Cerami A, Wolff SM, Kufe DW, Wilmore DW. Tumor necrosis factor and endotoxin induce similar metabolic responses in human beings. Surgery 1988; 104:280-286.

31. Duncombe AS, Brenner MK. Is circulating tumor necrosis factor bioactive? New England J Med 1988; 319:1227.

32. Fomsgaard A, Worsaae H, Bendtzen K. Detection of tumor necrosis factor from lipopolysaccharide-stimulated human mononuclear cells by enzyme-linked immunosorbent assay and cytotoxicity bioassay. Scand J Immunol 1988; 27:143-147.

33. Petersen CM, Moller BK. Immunological reactivity and bioactivity of tumour necrosis factor. The Lancet 1988; 1:934-935.

34. Shih ST, Khorram O, Lipton JM, McCann SM. Central administration of α-MSH antiserum augments fever in the rabbit. Am J Physiol 1986; 250:R803-R806.

35. Kasting NW. Criteria for establishing a physiological role for brain peptides. A case in point: the role of vasopressin in thermoregulation during fever and antipyresis. Brain Research Reviews 1989; 14:143-153.

36. Morrow LE, McClellan JL, Conn CA, Kluger MJ. Glucocorticoids alter

fever and IL-6 responses to psychological stress and to lipopolysaccharide. Am J Physiol 1993; 264:R1010-1016.

37. Jenkins JK, Malyak M, Arend WP. The effects of interleukin-10 on interleukin-1 receptor antagonist and interleukin-1β production in human monocytes and neutrophils. Lymphokine and Cytokine Research 1994; 13:47-54.

38. Ishida H, Muchamuel T, Sakaguchi S, Andrade S, Menon S, Howard M. Continuous administration of anti-interleukin 10 antibodies delays onset of autoimmunity in NZB/W F1 mice. J Exp Med 1994; 179:305-310.

39. Wang P, Wu P, Siegel MI, Egan RW, Billah MM. IL-10 inhibits transcription of cytokine genes in human peripheral blood mononuclear cells. J Immunol 1994; 153:811-816.

40. Zabel P, Schade FU, Schlaak M. Inhibition of endogenous TNF formation by pentoxifylline. Immunobiol 1993; 187:447-463.

41. LeMay LG, Vander AJ, Kluger MJ. The effects of pentoxifylline on lipopolysaccharide (LPS) fever and plasma interleukin-6 (IL-6) and tumor necrosis factor (TNF) in the rat. Cytokine 1990; 2:300-306.

42. Long NC, Kunkel SL, Vander AJ, Kluger MJ. Antiserum against TNF enhances LPS fever in the rat. Am J Physiol 1990; 258:R591-R595.

43. Smith BK, Kluger MJ. Anti-TNF-α antibodies normalized body temperature and enhanced food intake in tumor-bearing rats. Am J Physiol 1993; 265:R615-R619.

44. Kozak W, Conn CA, Wong GHW, Klir JJ, Kluger MJ. TNF soluble receptor and antiserum against TNF enhance lipopolysaccharide fever in mice. Am J Physiol 1995; 269:R23-R29.

45. Derijk RH, Berkenbosch F. Hypothermia to endotoxin involves the cytokine tumor necrosis factor and the neuropeptide vasopressin in rats. Am J Physiol 1994; 266:R9-R14.

46. Klir JJ, McClellan JL, Kozak W, Szelenyi Z, Wong GHW, Kluger MJ. Systemic but not central administration of tumor necrosis factor α attenuates LPS-induced fever in rats. Am J Physiol 1995; 268:R480-R486.

47. Long NC, Morimoto A, Nakamori T, Murakami N. Systemic injection of TNF-α attenuates fever due to IL-1β and LPS in rats. Am J Physiol 1992; 263: R987-R991.

48. Nagai M, Saigusa T, Shimada Y, Inagawa H, Oshima H, Iriki M. Antibody to tumor necrosis factor (TNF) reduces endotoxin fever. Experientia 1988; 44:606-607.

49. Kawasaki H, Moriyama M, Ohtani Y, Naitoh M, Tanaka A, Nariuchi H. Analysis of endotoxin fever in rabbits by using a monoclonal antibody to tumor necrosis factor (cachectin). Infect Immun 1989; 57:3131-3135.

50. Kwiatkowski D, Molyneux ME, Stephens S, Curtis N, Klein N, Pointaire P, Smit M, Allan R, Brewster DR, Grau GE, Greenwood BM. Anti-TNF therapy inhibits fever in cerebral malaria. Quart J Med 1993; 86:91-98.

51. Mathison JC, Wolfson E, Ulevitch RJ. Participation of tumor necrosis factor in the mediation of gram negative bacterial lipopolysaccharide-induced injury in rabbits. J Clin Invest 1988; 81:1925-1937.

52. Ebisui O, Fukata J, Murakami N, Kobayashi H, Segawa H, Muro S, Hanaoka I, Naito Y, Masui Y, Ohoto Y, Imura H, Nakao K. Effect of IL-1 receptor antagonist and antiserum to TNF-α on LPS-induced plasma ACTH and corticosterone rise in rats. Am J Physiol 1994; 266:E986-92.

53. McClellan JL, Klir JJ, Morrow LE, Kluger MJ. The central effects of glucocorticoid receptor antagonist RU38486 on lipopolysaccharide and stress-induced fever. Am J Physiol 1994; 267:R705-R711.

54. Leon LR, Kozak W, Peschon J, Kluger MJ. Exacerbated febrile responses to LPS, but not turpentine, in TNF double receptor knock-out mice. Am J Physiol 1996; In press.

55. Cooper AL, Brouwer S, Turnbull AV, Luheshi GN, Hopkins SJ, Kunkel SL, Rothwell NJ. Tumor necrosis factorα and fever after peripheral inflammation in the rat. Am J Physiol 1994; 267:R1431-R1436.

56. Sekut L, Menius JA Jr., Brackeen MF, Connolly KM. Evaluation of the significance of elevated levels of systemic and localized tumor necrosis factor in different animal models of inflammation. J Lab Clin Med 1994; 124:813-820.

57. Kapas L, Hong L, Cady AB, Opp MR, Postlethwaite AE, Seyer JM, Krueger JM. Somnogenic, pyrogenic, and anorectic activities of tumor necrosis factorα and TNF-α fragments. Am J Physiol 1992; 263:R708-15.

58. Kettelhut IC, Goldberg AL. Tumor necrosis factor can induce fever in rats without activating protein breakdown in muscle or lipolysis in adipose tissue. J Clin Invest 1988; 81:1384-1389.

59. Bluthé RM, Walter V, Parnet P, Layé S, Lestage J, Verrier D, Poole S, Stenning BE, Kelley KW, Dantzer R. Lipopolysaccharide induces sickness behaviour in rats by a vagal mediated mechanism. C R Adad Sci Paris, Sciences de la vie 1994; 317:499-503.

60. Goehler LE, Busch CR, Tartaglia N, Relton J, Sisk D, Maier SF, Watkins LR. Blockade of cytokine induced conditioned taste aversion by subdiaphragmatic vagotomy: further evidence for vagal mediation of immune-brain communication. Neuroscience Letters 1995; 185:163-166.

61. Watkins LR, Goehler LE, Relton JK, Tartaglia N, Silbert L, Martin D, Maier SF. Blockade of interleukin-1 induced hyperthermia by subdiaphragmatic vagotomy: evidence for vagal mediation of immune-brain commmunication. Neuroscience Letters 1995; 183:27-31.

62. Layé S, Bluthé RM, Kent S, Combe C, Médina C, Parnet P, Kelley K, Dantzer R. Subdiaphragmatic vagotomy blocks induction of IL-1β mRNA in mice brain in response to peripheral LPS. Am J Physiol 1995; 268:R1327-R1331.

63. Liao JF, Keiser JA, Scales WE, Kunkel SL, Kluger MJ. Role of corticosterone in TNF and IL-6 production in isolated perfused rat liver (IPRL). Am J Physiol 1995,268:R699-R706.

CHAPTER 6

CYTOKINE EFFECTS ON NEUROENDOCRINE AXES: INFLUENCE OF NITRIC OXIDE AND CARBON MONOXIDE

Andrew V. Turnbull and Catherine Rivier

IMMUNE-NEUROENDOCRINE INTERACTION AS A MECHANISM OF HOST DEFENSE

The existence of bilateral communication between the immune and neuroendocrine systems is now well established. The products of the neuroendocrine system, hormones such as corticosteroids, prolactin and growth hormone, have pronounced effects on a variety of aspects of immunoregulation, including T lymphocyte selection, splenic lymphocyte release and the expression and secretion of the intercellular mediators within the immune system (i.e., cytokines). Conversely, a number of cytokines exert potent effects on the synthesis and secretion of many hormones. Not surprisingly therefore, the immune and neuroendocrine systems share many of the same ligands and receptors, and function utilizing a common chemical language.[1] For example, lymphocytes contain mediators which are similar/identical to hormones, such as pro-opiomelanacortin products [adrenocorticotropin (ACTH), endorphins, enkephalins], growth hormone and prolactin, as well as expressing their respective receptors. On the other hand, the master endocrine gland, the pituitary, secretes and possesses cognate receptors for, a number of cytokines [e.g., interleukin-1 (IL-1), IL-6 and tumor necrosis factor-α (TNF-α)].

Such an intimate interaction between immunoregulatory and neuroendocrine components results in a fine tuning of each system

Cytokines in the Nervous System, edited by Nancy J. Rothwell. © 1996 R.G. Landes Company.

that is most apparent during (though not exclusive to) times of injury, infection, inflammation and/or disease. This is best exemplified
by considering the hypothalamo-pituitary (HP) axis response to these
threats. Activation of the HP-adrenal (HPA) and suppression of the
HP-gonadal (HPG) axes are two endocrine hallmarks of the body's
response to stress. Both of these responses to infection/disease are produced by the actions of cytokines elaborated during the accompanying
immune response. The resulting increase in corticosteroids (HPA response) participates in the regulation of a variety of metabolic pathways and cardiovascular responses. Corticosteroids also exert a restraining
influence on the ongoing immune response, and inadequate HPA responses are associated with an exacerbated inflammatory reactions and
increased susceptibility to inflammatory disease.[2] The suppression of
reproductive functions which occurs during injury/disease is necessary
to facilitate energy conservation, and to prevent unhealthy individuals
from reproducing. Indeed, stress appears to cause a shift in steroidogenic
pathways resulting in reduced sex steroid production and enhanced
synthesis of the corticosteroids necessary for vital metabolic reactions.
The mechanisms of this intimate association between immune and
neuroendocrine systems during host defense therefore underpin our
understanding of how the body mounts an appropriate (or inappropriate) response to stress, sickness and disease.

The purpose of this chapter is describe what we know about the
mechanisms of immune-neuroendocrine interaction from a standpoint
of the effects of cytokines on neuroendocrine axes, the major components of which are listed in Table 6.1. In particular, we will focus on
recent studies indicating that the novel messengers, nitric oxide and
carbon monoxide play, key roles in the regulation of cytokine-induced
neuroendocrine secretion.

CYTOKINES AND NEUROENDOCRINE SECRETION

CYTOKINE RECEPTORS WITHIN THE NEUROENDOCRINE SYSTEM

Cytokines exert their actions by interaction with specific, high affinity receptors, and are capable of eliciting cellular responses at very
low concentrations (in the picomolar range). Diverse actions of cytokines on neuroendocrine function are implied by a widespread distribution of their cognate receptors within neuroendocrine tissues.

The majority of work investigating the role of cytokines in the
regulation of neuroendocrine function has focused on the cytokines
IL-1, IL-6 and TNF-α. Receptors for these cytokines belong to three
distinct receptor families. Firstly, the immunoglobulin supergene family includes two identified IL-1 receptors, the 80 kDa type I receptor
and the smaller, 68 kDa type II receptor. Present evidence indicates
that type I receptors mediate the majority of IL-1 actions. Type II
receptors do not appear to transduce a cellular signal, but instead act

Table 6.1. Hormones of the neuroendocrine axes

Axis hormones	Hypothalamic releasing factor	Pituitary hormones	"Target organ"
		Adenohypophysial hormones	
HPA	corticotropin-releasing factor (CRF) (+)	adrenocorticotropin (ACTH)	corticosteroids
HPG	gonadotropin-releasing hormone (GnRH or LHRH) (+)	luteinizing hormone (LH) follicle-stimulating hormone (FSH)	testosterone progesterone estradiol
HPT	thyrotropin-releasing hormone (TRH) (+) somatostatin (SRIF) (-)	thyroid-stimulating hormone (TSH)	triiodothyronine (T_3) thyroxin (T_4)
somatotropic	growth hormone-releasing hormone (GRH) (+) somatostatin (-)	growth hormone (GH)	
	dopamine (-)	prolactin (PRL)	
		Neurohypophysial hormones	
		vasopressin oxytocin	

(+) = stimulatory (-) = inhibitory effects on respective pituitary hormone.

as a "receptor-decoy" which modifies cellular responsiveness to IL-1 by regulating the availability of IL-1 for interaction with the type I receptor.[3] A third member of the IL-1 receptor family has recently been identified and designated the IL-1 receptor accessory protein (IL-1R AcP).[4] This protein alone does not bind IL-1, but in the presence of the type I receptor, IL-1R AcP complexes with this receptor, and enhances the binding affinity of the type I receptor for IL-1. Secondly, the hematopoietic growth factor receptor family includes the receptors for IL-6 and related cytokines (oncostatin M, leukemia inhibitory factor, ciliary neurotropic factor and IL-11). Receptors belonging to this family are also multi-component in nature, consisting of a unique "receptor" (gp80 in the case of IL-6) and a signal transducing element (gp130), which is expressed ubiquitously. Finally, the TNF receptor family includes two receptors for both TNF-α and TNF-β (TNF-R60 and TNF-R80).

The distribution of these receptors within neuroendocrine tissues is indicated in Table 6.2. Due to the wide spectrum of actions of IL-1, IL-6 and TNF-α within the CNS, many studies have investigated the precise localization of receptors for these cytokines within the brain. While there is conflicting data regarding the localization of IL-1 receptors in rat brain, in general it appears that type I receptor mRNA[10,22,23] and binding of rat IL-1β[9] is mainly confined to non-neuronal cells. In particular, ependymal cells lining the ventricular system, choroid plexus, meninges and endothelial cells are major sites of IL-1 receptor expression. As such, it appears that IL-1 receptors (at least type I) are mainly expressed in "barrier-related" regions, and therefore provide a good anatomical substrate for the actions on the brain of IL-1 that is present within the bloodstream or within the cerebrospinal fluid. However, this is not a pertinent distribution for mediating the actions of IL-1 induced/expressed within the CNS parenchyma. Furthermore, the described localization of type I receptors and IL-1 binding in the rat brain are in stark contrast to those in the mouse where a dense population of type I IL-1 is detected in neuronal cells, particularly within the hippocampus.[5-8] These and other considerations (e.g., species differences in IL-1 binding,[24] heavy expression of IL-1R AcP in rat brain[4]) strongly suggest the existence of novel IL-1 receptors within the rat brain.

Far less work has focused on the distribution of IL-6 and TNF-α receptors within the CNS. Bovine hypothalamus exhibits binding of radiolabeled IL-6,[12] and IL-6 receptor mRNA is present in the preoptic, dorso-medial, and ventromedial areas of the rat hypothalamus, as well as in other brain regions (olfactory tract, piriform cortex and hippocampus).[11,25-28] With ^{125}I-labeled murine TNF-α as radioligand, either weak and diffuse[14] or specific (brainstem, thalamus, basal ganglia)[13] binding is present in mouse brain, and no binding is apparent in the rat.[14]

Table 6.2. Cytokine receptors in neuroendocrine tissues

Organ	Receptor	Species	Binding/mRNA	Example references
Brain	IL-1	Mouse	Binding	5,6
			mRNA	7,8
		Rat	Binding	9
			mRNA	10
	IL-6	Rat	mRNA	11
		Cow	Binding	12
	TNF-α	Mouse	Binding	13,14
		Rat	Binding	14
Pituitary	IL-1	Mouse	mRNA	7
			Binding	6
		Rat	Binding	9
	IL-6	Human	Binding	15
		Rat	Binding	15
			mRNA	15
	TNF-α	Mouse	Binding	14
		Rat	Binding	14
Testes	IL-1	Mouse	Binding	16
			mRNA	17
	IL-6	Rat	mRNA	18
Ovary	IL-1	Human	mRNA	19
Thyroid	IL-1	Human	Binding	20
		Rat	Binding	20
		Pig	Binding	21

EFFECTS OF CYTOKINES ON NEUROENDOCRINE AXES

The effects of cytokines on neuroendocrine axes has been investigated extensively, using both in vivo and in vitro systems. IL-1, IL-6 and TNF-α are known to activate the HPA axis in response to a variety of stressful stimuli, including endotoxemia, local inflammation, CNS viral disease and psychological stress (reviewed in ref. 29). In contrast, IL-1 appears to be an important inhibitory mediator of the HPG[30] and somatotropic (growth hormone)[31] axes responses to systemic administration of endotoxin.

Table 6.3 lists the effects of IL-1 on neuroendocrine secretion. Although numerous in vitro studies indicate that cytokines influence the secretion of hormones from the hypothalamus, pituitary and target endocrine gland, the site at which cytokines exert their primary effects in vivo depends on the specific axis under consideration, and on the source of cytokines. For example, the primary site of action of

Table 6.3. Effects of IL-1 on neuroendocrine secretion

Hormone	Secretion	Example References
Hypothalamic Releasing Factors		
CRF	↑	32-34
GnRH / LHRH	↓	35,36
GHRH	↓	31
SRIF	↑	31,37,38
TRH	↓	37,39
Anterior Pituitary Hormones		
ACTH	↑	29,40
LH	↓	30,41
FSH	↓	35
prolactin	↓ and ↑	42
TSH	↓	40,43
growth hormone	↓ and ↑	44,45
Posterior Pituitary Hormones		
oxytocin	↑	46
vasopressin	↑	46
Target Gland Hormones		
corticosteroids	↑	29
testosterone	↓	30,47
progesterone	↑	48
triiodothyronine (T_3) / thyroxin (T_4)	↑	37,40,43

IL-1, IL-6 and TNF-α in the induction of ACTH and corticosteroid secretion following administration of these cytokines into the bloodstream is at the level of hypophysiotropic nerve terminals in the median eminence, where they stimulate the secretion of corticotropin-releasing factor (CRF) (reviewed in ref. 29). When cytokines are injected directly into the brain, however, their influence appears to be primarily mediated at CRF cell bodies within the parvocellular region of the paraventricular nucleus (PVN) of the hypothalamus.[29] Considering the HPG axis, peripheral injection of IL-1, IL-6 and TNF-α decreases plasma testosterone concentrations, apparently without affecting gonadotropin-releasing hormone (GnRH) or luteinizing hormone (LH) secretion. This influence is exerted directly on the gonads via inhibition of steroidogenesis. In contrast, IL-1 injected into brain elicits decreases in secretion of all hormones of the axis primarily by inhibiting the activity of GnRH neurons within the medial pre-optic area of the hypothalamus (MPOA).[30,49]

Other neuroendocrine axes have been less well studied. Cytokines injected into the periphery decrease plasma thyroid stimulating hormone (TSH), thyroxin (T$_4$) and triiodothyronine (T$_3$) concentrations, reduce the plasma T$_3$ and T$_4$ responses to TSH, and incubation of hypothalamic explants with cytokines modulates the secretion of so-matostatin (a TSH-release inhibiting peptide) and thyrotropin-releas-ing hormone (TRH).[40] Both stimulatory and inhibitory effects of cen-trally injected IL-1 on plasma GH levels have been observed,[44,45] while peripherally administered IL-1 inhibits GH secretion.[45] The dual ef-fects of central IL-1 on GH secretion have been demonstrated to be related to dose,[44] and it is interesting to note that GH secretion is first inhibited, then exacerbated in response to bacterial lipopolysac-charide (LPS), both effects being prevented by treatment with IL-1 receptor antagonist.[31] IL-1 also decreases the secretion from hypotha-lamic explants of the GH-releasing hormone, GHRH, and inhibits the secretion of the GH-inhibiting hormone, somatostatin.[31] As with the somatotropic axis, plasma prolactin concentration responses to centrally injected-IL-1β are biphasic.[42] Acutely (within one hour), prolactin lev-els in intact male and female rats are reduced, but in females this is followed (by eight hours) by a marked and sustained elevation in pro-lactin secretion.[42] IL-1α injected centrally produces a marked increase in plasma prolactin levels within two hours in castrate male rats.[50] Cytokines also influence the secretion of neurohypophysial hormones, with significant increases in vasopressin and oxytocin concentrations being observed within both brain and blood.[46]

SOURCES OF CYTOKINES TO INFLUENCE NEUROENDOCRINE SECRETIONS

Increases in blood concentrations of cytokines are the most obvi-ous reason for increased exposure of neuroendocrine axes to these mediators. Such increases in concentrations are observed after viral, bacterial, hypoxic, traumatic and even psychological insults (see ref. 29 for detailed discussion). However, the temporal relationship be-tween the appearance of cytokines in blood and subsequent neuroen-docrine changes remains poorly understood. During discrete local in-flammation, however, the appearance of IL-6 in blood correlates well with the subsequent increase in HPA activity.[51,52] We find that plasma TNF-α concentrations increase before HPA activation due to intrave-nous endotoxin administration, though others report that plasma ACTH concentrations rise in response to intraarterial endotoxin before the appearance of either IL-1, IL-6 or TNF-α in blood.[53] However, to what extent increased blood levels of cytokines reflects a particular neu-roendocrine axes' exposure to that cytokine is unclear. Firstly, there has been little attempt to determine whether the blood concentrations measured during a particular inflammatory response are sufficient to induce the accompanying neuroendocrine event. Secondly, while in-creases in the blood concentration clearly result in increased exposure of endocrine glands such as the pituitary, adrenal, gonads or thyroid to a cytokine, neuroendocrine responses appear to be mediated by the hypothalamus. The blood-brain-barrier (BBB) is relatively imperme-able to cytokines (molecular weights 8-40 kDa), and while saturable transport mechanisms of IL-1,[54] IL-6[55] and TNF-α[56] have been de-scribed, these studies indicate that accumulation of these cytokines within the brain is only small (in general, < 0.5% of injected cytokine pen-etrates into brain parenchyma). Therefore substantial accumulation of cytokines in brain as a result of elevated blood concentrations would appear to occur only after large and sustained elevations. Alternatively, cytokines may act at areas of the CNS relatively devoid of a BBB, such as the OVLT and of particular interest to the neuroendocrine system–the median eminence.[57] Finally, their concentrations in blood may not be a good indicator of neuroendocrine exposure to cytokines because neuroendocrine tissues (including the brain) themselves syn-thesize and secrete a variety of these mediators. This raises the possi-bility that cytokines, which were originally described as intercellular mediators within the immune system, may also act in a paracrine fashion within neuroendocrine tissues.

A variety of cytokines (including interleukins, tumor necrosis fac-tors and growth factors) are synthesized within the CNS. Potential cellular sources include invading macrophages, endothelial cells, microglia, astrocytes as well as neurons themselves (see ref. 58 for a detailed re-view). Elevated synthesis or secretion of cytokines within the brain in response to CNS infection, hypoxia, injury, ischemia, convulsion or neurodegenerative disease have been reported. In addition, it has been suggested that infection or inflammation within the periphery may also lead to elevated cytokine synthesis within the brain. However, elevated IL-1, IL-6 and TNF-α synthesis within the brain due to peripheral stimuli have been reported in response to the administration of only extremely large amounts of endotoxin.[59-64] Smaller doses, which in our experience still produce pronounced neuroendocrine alterations, pro-duce either no measurable effect on CNS cytokine synthesis or in-creases which occur long after the initiation of the accompanying neu-roendocrine event. Therefore, what the relationship between CNS cytokine synthesis and neuroendocrine alterations in response to pe-ripheral infectious/inflammatory stimulis, remains an important ques-tion. Similarly, the nature and mechanisms of neuroendocrine responses to CNS pathologies, which unquestionably provoke substantial CNS cytokine synthesis and secretion, require further investigation.

There is now substantial evidence for the existence of cytokine networks within endocrine organs, such as the pituitary, adrenal, tes-tes and ovaries. Not only do they posses cytokine receptors (Table 6.2), but these organs also synthesize and secrete a number of cytokines (including IL-1, IL-6 and TNF-α) (see ref. 65 for references). Not surprisingly, the synthesis and secretion of IL-1, IL-6 and TNF-α by these tissues is induced by endotoxin. However, perhaps more interesting is that a variety of "classical" hormone signals also stimulate cytokine secretion in neuroendocrine tissues. For example, calcitonin gene-related peptide and vasoactive-intestinal peptide increase IL-6 secretion from rat anterior pituitary cells,[66] ACTH induces IL-6 secretion from rat adrenal zona glomerulosa cells[67] and gonadotropins stimulate IL-1 synthesis in the rat ovary,[68] suggesting a role for cytokines in neuroendocrine regulation, unrelated to sickness/disease/stress. Indeed, cytokines may influence the development and growth of neuroendocrine cells.[69]

INFLUENCE OF NITRIC OXIDE AND CARBON MONOXIDE ON NEUROENDOCRINE AXES

ROLE OF INTERMEDIATES IN CYTOKINE MODULATION OF NEUROENDOCRINE AXES

As indicated above, cytokine receptors are distributed within neuroendocrine tissues. However, there are only a few examples where the presence of these receptors has clearly been localized to hormone-producing cells (e.g., IL-1 receptors on the mouse corticotropic cell line, AtT-20,[70,71] IL-2 receptors on a variety of pituitary hormone-producing cells,[69] and IL-6 receptors on testicular Leydig cells[18]). Of particular note is the apparent lack of IL-1 receptors on neuroendocrine cells within the rat hypothalamus. Consequently, there has been considerable interest in the role of intermediary signals between cytokine actions and neuroendocrine secretion. The role of excitatory amino acids, opioid peptides, prostaglandins and catecholamines in mediating cytokine-induced modulation of hypothalamic neuroendocrine secretion has been reviewed extensively elsewhere.[30,72] However, the recent emergence of a new class of neuromodulators, namely the gaseous molecules nitric oxide and carbon monoxide,[73] has led to a number of studies investigating the role of these mediators in neuroendocrine function.

NITRIC OXIDE SYNTHASE IS PRESENT IN NEUROENDOCRINE TISSUES

Nitric oxide synthase (NOS) is the enzyme that catalyzes the conversion of L-arginine to L-citrulline and the gaseous mediator, nitric oxide (NO). Several isoforms of NOS have been identified (reviewed in ref. 74). NOS I is expressed in the central and peripheral nervous systems and is otherwise known as brain or neuronal NOS. NOS II is found in many cell types, such as hepatocytes, macrophages, smooth muscle cells and glia. Finally, NOS III is synonymous with endothelial NOS. These three isoforms are distinct gene products, differ in terms of their constitutive expression (NOS I and III are constitutively expressed, whereas NOS II is present only after induction by cytokines or endotoxin), are either calcium/calmodulin-dependent (NOS

I and III) or -independent (NOS II), and exhibit different kinetic properties. That NO acts as a neuromodulator was first indicated by the demonstration that inhibitors of NOS block the stimulation of cGMP synthesis in brain slices by glutamate acting at NMDA receptors. NO clearly does not behave like a conventional neurotransmitter, since it is neither stored in nerve terminals nor does it influence its target cell via interaction with a cell-surface receptor. Rather it diffuses from nerve terminals and forms covalent linkages with several potential targets (e.g., guanyl cyclase).

NOS-like activity, NOS immunoreactivity and NOS mRNA are present within the PVN and SON of the hypothalamus.[75-86] Within the magnocellular division of the PVN, NOS is upregulated by salt loading[84] and, at least in some neurons, is colocalized with vasopressin,[77] data indicative of a role for NO in the neuroendocrine regulation of body fluid balance. NO is also present within the parvocellular division of the PVN, and is present in a subpopulation of CRF-expressing neurons.[81] Circumstances known to alter the neuroendocrine activity of the PVN, e.g., salt loading,[84] lactation,[79] endotoxin treatment[80] and immobilization stress,[78] upregulate NOS expression in this nucleus, suggesting a role for NO in the regulation of neuroendocrine function.

With regard to localization within the pituitary gland, NOS is present in both the posterior[76,86] and anterior[87] pituitary. Within the posterior pituitary, NOS presumably represents nerve endings originating from the magnocellular neurosecretory neurons. As for the anterior pituitary, NOS mRNA and protein is expressed at only low levels in normal animals. However, after castration a marked increase in NOS is apparent. Double immunostaining experiments indicate colocalization of NOS with LH, and in addition, a second population of NOS-positive cells (folliculo-stellate cells) that are in close proximity to somatotrophs (GH-producing cells).[87] Apart from the hypothalamus and pituitary, NOS is also present in the hypophysial portal vasculature,[88] which delivers hypothalamic releasing factors to the anterior pituitary, and within the reproductive organs[89] and the adrenal medulla.[90,91]

NITRIC OXIDE INFLUENCES NEUROENDOCRINE SECRETIONS

Possible roles of nitric oxide in the regulation of neuroendocrine secretion have been determined using NO donors (L-arginine, nitroprusside), inhibitors of NOS activity (arginine derivatives), and NO scavengers (hemaglobin). NOS inhibitors are generally low molecular weight (150-300 D) and the BBB does not appear to pose much of a hindrance to their diffusion. These inhibitors have therefore been studied after either systemic or central injection, the latter being chosen when effects on systemic parameters, in particular blood pressure, were to be avoided. Injection of N[G]-monomethyl-L-arginine (L-NMMA) into the third ventricle of castrate male rats blocks the pulsatile secretion of LH, but not FSH,[92] while peripheral injection of N^G-methyl-L-arginine (L-NNA) or Nω-nitro-L-arginine methyl ester (L-NAME) inhibit the steroid-induced surge of LH release in ovariectomized female rats.[93] NO donors (e.g., sodium nitroprusside) increase LHRH release from hypothalamic explants,[92] and the increased LHRH secretion induced by excitatory amino acids[94] and norepinephrine[92] are prevented by L-NMMA, suggesting that NO plays a significant role in the regulation of LHRH secretion. However, the intrahypothalamic actions of NO on LHRH secretion appears to be complex, since NO inhibits LHRH release in response to γ-amino-butyric acid.[95] Similarly, both inhibitory and stimulatory roles have been suggested for NO in the regulation of prolactin secretion. NOS inhibitors prevent dopamine-induced *reductions* in prolactin secretion[96] from anterior pituitary cell cultures, while they blunt IL-1α-induced *elevations*[50] and steroid-induced surges[97] in plasma prolactin concentrations. Using hypothalamic or pituitary cell cultures and/or in vivo protocols, NO has been implicated in the regulation of the HPA,[98-104] HPT[105] and somatotropic axes,[106,107] as well as the release of the neurohypophysial hormones, oxytocin[108] and vasopressin.[109,110]

ROLE OF NITRIC OXIDE IN THE REGULATION OF HPA AXIS RESPONSE TO CYTOKINES

NO is a key mediator of inflammatory processes. It causes vasodilation, increased vascular permeability, increased leukocyte adhesion and is responsible for microbial killing.[111] In addition to these actions, there is substantial anecdotal evidence that NO may play a significant role in the regulation of HPA responses to inflammation. Firstly, NOS is present in both hypothalamic nuclei and pituitary regions relevant to the control of the HPA axis. Furthermore, NO appears to be intimately related to the control of neuroendocrine function. Finally, the actions of cytokines on neuroendocrine axes are subject to the influence of intermediary signals, and the expression of at least one isoform of NOS is cytokine-inducible.[112] We have conducted a number of in vivo studies investigating the role of NO in the regulation of the HPA axis, in particular the HPA responses to the systemic (intravenous, iv) injection of low doses of recombinant IL-1β (25-400 ng/kg), and our results have provided strong evidence that NO plays an important role in restraining HPA responses to inflammatory stimuli.[52,98,100,113]

Initial experiments demonstrated that iv pretreatment of rats with the NOS substrate L-arginine, blunted the rise in plasma ACTH due to iv IL-1β. Conversely the NOS inhibitor, L-NAME (iv), exaggerates and prolongs the ACTH and corticosterone responses to IL-1β (Fig. 6.1), an action which is specific to the L-isomer of NAME (the isomer active at NOS) and can be reversed by competition with L-arginine (the substrate for NOS). Furthermore, the potentiation of IL-1-induced ACTH secretion produced by L-NAME appears to be unrelated to its

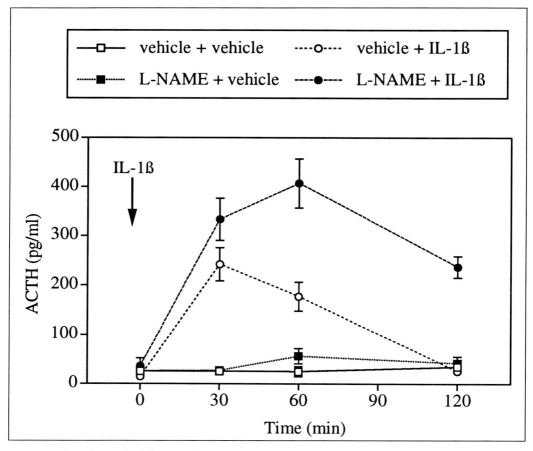

Fig. 6.1. The effect of inhibition of nitric oxide synthase (NOS) on the increase in plasma ACTH concentration due to IL-1β in adult (70 days old), male rats. The NOS inhibitor, L-NAME (30 mg/kg), or its vehicle, was administered intravenously (iv) two minutes before the iv injection of 400 ng/kg IL-1β. L-NAME produced a significant (P < 0.01) potentiation and prolongation of the ACTH response to IL-1β.

hypertensive effects. Indeed L-NAME produces similar increases in blood pressure in both vehicle and IL-1β-treated rats, but results in marked elevations in ACTH secretion only in the latter group. L-NAME similarly potentiates and prolongs ACTH secretion due to IL-1β in pre-, peri- and post-pubertal, male and female rats (Fig. 6.2). This indicates that the influence of NO is independent of the sex steroid meilieu, and suggests that this NOergic mechanism is present early in postnatal endocrine development. Potentiation of ACTH secretion by L-NAME is also observed during responses to local inflammation (turpentine-induced tissue damage) and systemic endotoxin treatment, but interestingly, not to IL-1β administered directly into the brain.

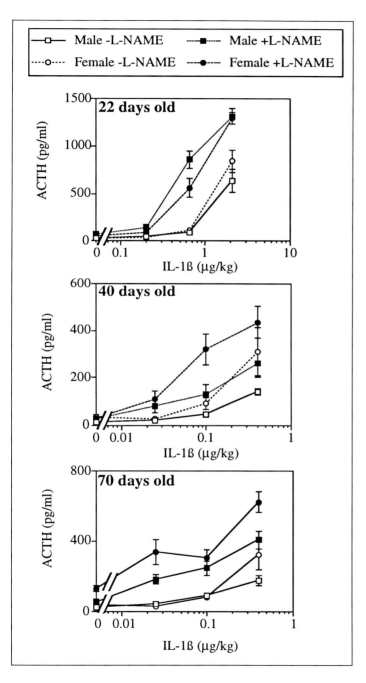

Fig. 6.2. The effect of inhibition of NOS on the ACTH response to IL-1β in pre- (22 days old), peri- (40 days old) and post- (70 days old) pubertal, male and female rats. The NOS inhibitor, L-NAME (30 mg/kg), was administered either subcutaneously (sc, 22 days old) or intravenously (iv, 40 and 70 days old), before the injection of IL-1β (sc - 22 days old; iv - 40 and 70 days old). The data presented are the plasma ACTH concentrations two hours (22 days old) or one hour (40 and 70 days old) after IL-1β. For both males and females, at all ages, L-NAME caused a parallel leftward shift in the dose-response curves, indicating an increased responsiveness to IL-1β.

The mechanisms by which NO influences the HPA response to IL-1β, inflammation and endotoxemia are far from clear, but a number of findings have shed some light on likely candidates. Since CRF is an obligatory mediator of the ACTH response to IL-1, a number of in vitro studies have investigated whether NO influences hypothalamic CRF secretion. However, in hypothalamic explants NOS inhibitors either blunt[104,114] or exacerbate[101] the CRF response to IL-1. Similarly, L-NNA either blunts[114] or has no effect[115] on IL-1-induced ACTH secretion in cultures of rat anterior pituitaries. We have found that in vivo, L-NAME does not influence ACTH secretion due to CRF, but potentiates the ACTH response to the two less potent ACTH secretagogs, vasopressin and oxytocin.[98] These latter findings raise the possibility that increased sensitivity of corticotropes to vasopressin and/or oxytocin following L-NAME accounts for its potentiation of IL-1β-induced ACTH secretion. However, administration of neutralizing antisera to vasopressin does not influence the exacerbation of IL-1 induced ACTH secretion produced by L-NAME.[100] On the other hand, administration of combined adrenergic antagonists, propanolol and prazosin, partially reverse the effects of L-NAME on IL-1-, but not vasopressin-induced ACTH secretion.[100] These data indicate that unlike in normal subjects, in animals in which NOS has been inhibited, ACTH secretion due to low doses of IL-1 is at least partially mediated by catecholamines, and suggests that the inhibitory influence of NO on HPA activity is due to a suppressive effect on catecholaminergic pathways.

CARBON MONOXIDE

In general, neurotransmitters belong to chemical classes, such as amino acids, peptides, amines. Therefore it is not surprising that NO does not appear to be the only gaseous neuromodulator. Carbon monoxide (CO) shares many of the biological properties of NO, including activation of guanyl cyclase and consequent dilation of blood vessels.[73] CO is generated in the body by heme oxygenase (HO), an enzyme which cleaves the heme ring into biliverdin and CO. As is the case for NO, heme oxygenase exists in both inducible (HO-1) and constitutive (HO-2) isoforms. HO-1 is induced by a variety of exogenous and endogenous chemicals, including heme itself, endotoxins and by oxidative stressors. It is found in all organs (particularly the spleen and liver) but is expressed at very low levels in brain.[116] HO-1 is responsible for the destruction of heme from senescent red blood cells. HO-2 is abundant within the brain,[116] indicating a role for HO-2 outside of destruction of heme, since this organ is thought not to play a major role in heme turnover. The colocalization of HO-2 and guanyl cyclase mRNA within brain suggests a physiological role for this enzyme in neurotransmission.[117] Indeed, inhibition of HO with protoporphyrin-IX (PP-IX) inhibits cGMP formation in primary cultures of olfactory neurons[117] and reverses long-term potentiation in pyramidal cell slices.[118]

While HO-2 mRNA has been demonstrated in the rat hypothalamus,[119] at present only a few studies have addressed the possible role of CO in neuroendocrine regulation. In hypothalamic explants, a substrate of HO, hematin, inhibits CRF release and reverses IL-1-induced CRF secretion.[120] However, in primary hypothalamic cultures, hematin stimulates, while PP-IX inhibits, basal CRF secretion.[121] In hypothalamic fragments, hematin markedly stimulates GnRH secretion, an effect blocked by PP-IX and hemaglobin (a CO scavenger) and not mimicked by biliverdin, clearly demonstrating that the observed effects of hematin are due to CO produced by HO.[122] We have recently obtained some preliminary in vivo data, indicating that CO, like NO, restrains the HPA response to IL-1β. Figure 6.3 shows that pretreatment of rats with PP-IX dramatically potentiates the elevation in plasma ACTH concentrations induced by iv IL-1.

CONCLUSIONS

The present chapter has described the effects of cytokines on neuroendocrine function, and discussed some of the potential mechanisms

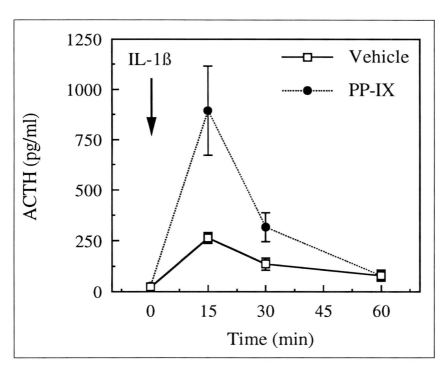

Fig. 6.3. The effect of inhibition of heme oxygenase (HO) on the ACTH response to 100 ng/kg IL-1β in 70 day old male rats. Protoporphyrin-IX (PP-IX, 50 mg/kg) was administered 20 and 3 h before IL-1β. PP-IX had no effect on basal concentrations of ACTH, but produced a significant (P < 0.05) enhancement of the ACTH response to IL-1β.

for these interactions. That immune and neuroendocrine systems utilize a common chemical language is becoming increasingly apparent; each manufacturing similar ligands and receptors. The data discussed in the latter half of this chapter indicates that the gaseous mediator nitric oxide (and possibly carbon monoxide) represent a new class of mediators which are generated by, and influence the activity of, both immune and neuroendocrine systems.

ACKNOWLEDGMENTS

We are grateful to Dr. Tony Troutt of Immunex for the generous gift of recombinant IL-1β. Portions of the work described in this chapter were supported by NIH Grant MH-51774, the Foundation for Research Inc. (CR), and Aaron and Amoco fellowships (AVT).

REFERENCES

1. Blalock JE. A molecular basis for bidirectional communication between the immune and neuroendocrine systems. Physiol Rev 1989; 69:1-32.
2. Chrousos GP. The hypothalamic-pituitary-adrenal axis and immune-mediated inflammation. N Engl J Med 1995; 332:1351-62.
3. Sims JE, Giri JG, Dower SK. The two interleukin-1 receptors play different roles in IL-1 actions. Clin Immunol Immunopathol 1994; 72:9-14.
4. Greenfeder SA, Nunes P, Kwee L et al. Molecular cloning and characterization of a second subunit of the interleukin 1 receptor complex. J Biol Chem 1995; 270:13757-65.
5. Ban EM, Milon G, Fillion G et al. Receptors for interleukin-1 (α and β) in mouse brain: mapping and neuronal localization in hippocampus. Neuroscience 1991; 43:21-30.
6. Ban E, Marquette C, Sarrieau A et al. Regulation of interleukin-1 receptor expression in mouse brain and pituitary by lipopolysaccharide and glucocorticoids. Neuroendocrinology 1993; 58:581-87.
7. Cunningham ET Jr, Wada E, Carter DB et al. In situ histochemical localization of type I interleukin-1 receptor mRNA in the central nervous system, pituitary, and adrenal gland of the mouse. J Neurosci 1992; 12:1101-14.
8. Cunningham ET Jr, Wada E, Carter DB et al. Localization of interleukin-1 receptor messenger RNA in murine hippocampus. Endocrinology 1991; 128:2666-68.
9. Marquetee C, Van Dam A-M, Can E et al. Rat interleukin-1β binding sites in rat hypothalamus and pituitary gland. Neuroendocrinology 1995; 62:362-69.
10. Ericsson A, Liu C, Hart RP et al. Type I interleukin-1 receptor in the rat brain: distribution, regulation and relationship to sites in IL-1-induced cellular activation. J Comp Neurol 1995; 361:681-98.
11. Schobitz B, Voorhuis DAM, De Kloet ER. Localization of interleukin-6 mRNA and interleukin 6 receptor mRNA in rat brain. Neurosci Lett 1992; 136:189-92.

12. Cornfield LJ, Sills MA. High affinity interleukin-6 binding sites in bovine hypothalamus. Eur J Pharmacol 1991; 202:113-15.

13. Kinouchi K, Brown G, Pasternak G et al. Identification and characterization of receptors for tumor necrosis factor-alpha in the brain. Biochem Biophys Res Com 1991; 18:1532-38.

14. Wolvers DA, Marquette C, Berkenbosch F et al. Tumor necrosis factor-alpha: specific binding sites in rodent brain and pituitary gland. Eur Cytokine. Net 1993; 4:377-81.

15. Ohmichi M, Hirota K, Koike K et al. Binding sites for interleukin-6 in the anterior pituitary gland. Neuroendocrinology 1992; 55:199-203.

16. Takao T, Mitchell WM, Tracey DE et al. Identification of interleukin-1 receptors in mouse testis. Endocrinology 1990; 127:251-58.

17. Cunningham ET Jr, Wada E, Carter DB et al. Distribution of Type I interleukin-1 receptor mRNA in testis: an in situ histochemical study in the mouse. Neuroendocrinology 1992; 56:94-99.

18. Okuda Y, Morris PL. Identification of interleukin-6 receptor (IL-6R) mRNA in isolated Sertoli and Leydig cells: regulation by gonadotropin and interleukins in vitro. Endocrine 1994; 2:1163-68.

19. Hurwitz A, Loukides J, Ricciarelli E et al. Human intraovarian interleukin-1 (IL-1) system: highly compartmentalized and hormonally dependent regulation of the genes encoding IL-1, its receptor, and its receptor antagonist. J Clin Invest 1992; 89:1746-54.

20. Svenson M, Kayser L, Hansen MB et al. Interleukin-1 receptors on human thyroid cells and on the rat thyroid cell line FRTL-5. Cytokine 1991; 3:125-30.

21. Kasai K, Hiraiwa M, Emoto T et al. Presence of high affinity receptor for interleukin-1 (IL-1) on cultured porcine thyroid cells. Horm Metabol Res 1990; 22:75-79.

22. Yabuuchi K, Minami M, Katsumata S et al. Localization of type I interleukin-1 receptor mRNA in the rat brain. Mol Brain Res 1994; 27:27-36.

23. Wong M-L, Licinio J Localization of interleukin 1 type I receptor mRNA in rat brain. Neuroimmunomodulation 1994; 1:110-15.

24. Takao T, Newton RC, De Souza EB. Species differences [125I]interleukin-1 binding in brain, endocrine and immune tissues. Brain Res 1993; 623:172-76.

25. Schobitz B, De Kloet ER, Sutanto W et al. Cellular localization of interleukin-6 mRNA and interleukin-6 receptor mRNA in rat brain. Eur J Neurosci 1993; 5:1426-35.

26. Gradient RA, Otten U. Differential expression of interleukin-6 (IL-6) and interleukin-6 receptor (IL-6R) mRNAs in rat hypothalamus. Neurosci Lett 1993; 153:13-16.

27. Gradient RA, Otten U. Expression of interleukin-6 (IL-6) and interleukin-6 receptor (IL-6R) mRNAs in rat brain during postnatal development. Brain Res 1994; 637:10-14.

28. Gradient RA, Otten U. Identification of interleukin-6 (IL-6)-expressing

neurons in the cerebellum and hippocampus of normal adult rats. Neurosci Lett 1994; 182:243-46.

29. Turnbull AV, Rivier CL. Regulation of the HPA axis by cytokines. Brain Behav Immun 1995; 9:253-75.

30. Rivest S, Rivier C. The role of corticotropin-releasing factor and interleukin-1 in the regulation of neurons controlling reproductive functions. Endocr Rev 1995; 16:177-99.

31. Peisen JN, McDonnell KJ, Mulroney SE et al. Endotoxin-induced suppression of the somatotropic axis is mediated by interleukin-1β and corticotropin-releasing factor in the juvenille rat. Endocrinology 1995; 136:3378-90.

32. Sapolsky R, Rivier C, Yamamoto G et al. Interleukin-1 stimulates the secretion   of hypothalamic corticotropin-releasing factor. Science 1987; 238:522-24.

33. Watanobe H, Takebe K. Effects of intravenous administration of interleukin-1 beta on the release of prostaglandin E2, corticotropin-releasing factor, and arginine vasopressin in several hypothalamic areas of freely moving rats: estimation by push-pull perfusion. Neuroendocrinology 1994; 60:8-15.

34. Tsagarakis S, Gillies G, Rees LH et al. Interleukin-1 directly stimulates the release of corticotrophin releasing factor from rat hypothalamus. Neuroendocrinology 1989; 49:98-101.

35. Rivest S, Lee S, Attardi B et al. The chronic intracerebroventricular infusion of interleukin-1β alters the activity of the hypothalamic-pituitary-gonadal axis of cycling rats. I. Effect on LHRH and gonadotropin biosynthesis and secretion. Endocrinology 1993; 133:2424-30.

36. Rivest S, Rivier C. Centrally injected interleukin-1β inhibits the hypothalamic LHRH secretion and circulating LH levels via prostaglandins in rats. J Neuroendocrinol. 1993; 5:445-50.

37. Kennedy JA, Wellby ML, Zotti R. Effect of interleukin-1β, tumor necrosis factor-α and interleukin-6 on the control of thyrotropin secretion. Life Sci 1995; 57:487-501.

38. Scarborough DE, Lee S, Dinarello CA et al. Interleukin-1β stimulates somatostain biosynthesis in primary culture of fetal rat brain. Endocrinology 1989; 124:549-51.

39. Pang X-P, Hershman JM, Mirell CJ et al. Impairment of hypothalamic-pituitary-thyroid function in rats treated with human recombinant tumor necrosis factor-α (cachectin). Endocrinology 1989; 125:76-84.

40. Imura H, Fukata J-I, Mori T. Cytokines and endocrine function: an interaction between the immune and neuroendocrine systems. Clin Endocrinology 1991; 35:107-15.

41. Rivier C, Vale W. Cytokines act within the brain to inhibit luteinizing hormone secretion and ovulation in the rat. Endocrinology 1990; 127:849-56.

42. Rivier C. Mechanisms of altered prolactin secretion due to the administration of interleukin-1β into the brain ventricles of the rat. Neuroendo-

crinology 1995; 62:198-206.

43. Stouthard JML, van der Poll T, Endert E et al. Effects of acute and chronic interleukin-6 administration on thyroid hormone metabolism in humans. J Clin Endocrinol Metab 1994; 79:1342-46.

44. Payne LC, Obal F, Opp MR et al. Stimulation and inhibition of growth hormone secretion by interleukin-1β: the involvement of growth hormone-releasing hormone. Neuroendocrinology 1992; 56:118-23.

45. Wada Y, Sato M, Niimi M et al. Inhibitory effect of interleukin-1 on growth hormone secretion in conscious male rats. Endocrinology 1995; 136:3936-41.

46. Landgraf R, Neumann I, Holsboer F et al. Interleukin-1β stimulates both central and peripheral release of vasopressin and oxytocin in the rat. Eur J Neurosci 1995; 7:592-98.

47. van der Poll T, Romijn JA, Endert E et al. Effects of tumor necrosis factor on the hypothalamic-pituitary-testicular axis in healthy men. Metabolism 1993; 42:303-07.

48. Rivier C, Erickson G. The chronic intracerebroventricular infusion of interleukin-1β alters the activity of the hypothalamic-pituitary-gonadal axis of cycling rats. II. Induction of pseudopregnant-like corpora lutea. Endocrinology 1993; 133:2431-36.

49. Turnbull A, Rivier C. Brain-periphery connections: do they play a role in mediating the effect of centrally injected interleukin-1β on gonadal function? Neuroimmunomodulation 1996; 2:224-235.

50. Rettori V, Belova N, Gimeno M et al. Inhibition of nitric oxide synthase in the hypothalamus blocks the increase in plasma prolactin induced by intraventricular injection of interleukin-1α in the rat. Neuroimmunomodulation 1994; 1:116-20.

51. Turnbull AV, Dow RC, Hopkins SJ et al. Mechanism of the activation of the pituitary-adrenal axis by tissue injury in the rat. Psychoneuroendocrinology 1994; 19:165-78.

52. Turnbull AV, Rivier C. CRF, vasopressin and prostaglandins mediate, and nitric oxide restrains, the HPA axis response to acute local inflammation in the rat. Endocrinology 1996; 137:455-63.

53. Givalois L, Dornand J, Mekaouche M et al. Temporal cascade of plasma level surges in ACTH, corticosterone, and cytokines in endotoxin-challenged rats. Am J Physiol 1994; 267:R164-70.

54. Banks WA, Ortiz L, Plotkin SR et al. Human interleukin (IL) -1α, murine IL-1α and murine IL-1β are transported from blood to brain in the mouse by a shared saturable transport mechanism. J Pharmacol Exp Ther 1991; 259:988-96.

55. Banks WA, Kastin AJ, Gutierrez EG. Penetration of interleukin-6 across the murine blood-brain-barrier. Neurosci Lett 1994; 179:53-56.

56. Gutierrez EG, Banks WA, Kastin AJ Murine tumor necrosis factor alpha is transported from blood to brain in the mouse. J Neuroimmunol 1993; 47:169-176.

57. Stitt JT. Passage of immunomodulators across the blood-brain barrier. Yale

J Biol Med 1990; 63:121-31.

58. Hopkins SJ, Rothwell NJ. Cytokines and the Nervous System I: expression and regulation. Trends Neurosci 1995; 18:83-88.

59. Ban EM, Haour F, Lenstra R. Brain interleukin-1 gene expression induced by peripheral lipopolysaccharide administration. Cytokine 1992; 4:48-54.

60. van Dam A-M, Brouns M, Louisse S et al. Appearance of interleukin-1 in macrophages and in ramified microglia in the brain of endotoxin-treated rats: a pathway for the induction of non-specific symptoms of sickness? Brain Res 1992; 588:291-96.

61. Gatti S, Bartfai T. Induction of tumor necrosis factor-alpha mRNA in the brain after peripheral endotoxin treatment—comparison with interleukin-1 family and interleukin-6. Brain Res 1993; 624:291-94.

62. Buttini M, Boddeke H. Peripheral lipopolysaccharide stimulation induces interleukin-1β messenger RNA in rat brain microglial cells. Neuroscience 1995; 65:523-30.

63. van Dam A-M, Bauer J, Tilders FJH et al. Endotoxin-induced appearance of immunoreactive interleukin-1β in ramified microglia in rat brain: a light and electron microscopic study. Neuroscience 1995; 65:815-26.

64. Hagan P, Poole S, Bristow AF. Endotoxin-stimulated production of rat hypothalamic interleukin-1β *in vivo* and *in vitro*, measured by specific immunoradiometric assay. J Mol Endocrinol 1993; 11:31-36.

65. Turnbull AV, Rivier C. Cytokines within the neuroendocrine system. Curr Opinion Endocrinol Diabetes 1996; 3:149-156.

66. Tatsuno I, Somogyvari-Vigh A, Mizuno K et al. Neuropeptide regulation of interleukin-6 production from the pituitary: stimulation by adenylate cyclase activating polypeptide and calcitonin gene-related peptide. Endocrinology 1991; 129:1797-804.

67. Judd AM, MacLeod RM. Differetial release of tumor necrosis factor and IL-6 from adrenal zona glomerulosa cells in vitro. Am J Physiol 1995; 268:E114-20.

68. Hurwitz A, Ricciarelli E, Botero L et al. Endocrine and autocrine-mediated regulation of rat ovarian (theca-interstitial) interleukin-1β gene expression: gonadotropin-dependent preovulatory acquisition. Endocrinology 1991; 129:3427-29.

69. Arzt E, Sauer J, Buric R et al. Characterization of interleukin-2 (IL-2) receptor expression and action of IL-1 and IL-6 on normal anterior pituitary cell growth. Endocrine 1995; 3:113-19.

70. Webster EL, Tracey DE, De Souza EB. Upregulation of interleukin-1 receptors in mouse AtT-20 pituitary cells following treatment with corticotropin-releasing factor. Endocrinology 1991; 129:2796-98.

71. Bristulf J, Bartfai T. Interleukin-1β and tumor necrosis factor-α stimulate the mRNA expression of interleukin-1 receptors in mouse anterior pituitary AtT-20 cells. Neurosci Lett 1995; 187:53-56.

72. Rivier C. Influence of immune signals on the hypothalamic-pituitary axis of the rodent. Front Neuroendocrinol 1995; 16:151-82.

73. Dawson TM, Snyder SH. Gases as biological messengers: nitric oxide and carbon monoxide in the brain. J Neurosci 1994; 14:5147-59.

74. Forstermann U, Gath I, Schwarz P et al. Isoforms of nitric oxide synthase. Properties, cellular distribution and expressional control. Biochem Pharmacol 1995; 50:1321-32.

75. Arevalo R, Sanchez F, Alonso JR et al. NADPH-diaphorase activity in the hypothalamic magnocellular neurosecretory nuclei of the rat. Brain Res Bull 1992; 28:599-603.

76. Bredt DS, Hwang PM, Snyder SH. Localization of nitric oxide synthase indicating a neural role for nitric oxide. Nature 1990; 347:768-70.

77. Calka J, Block CH. Relationship of vasopressin with NADPH-diaphorase in the hypothalamo-neurohypophysial system. Brain Res Bull 193; 32:207-10.

78. Calza L, Giardino L, Ceccatelli S. NOS mRNA in the paraventricular nucleus of young and aged rats after immobilization stress. Neuroreport 1993; 4:627-30.

79. Ceccatelli S, Eriksson M. The effect of lactation on nitric oxide synthase gene expression. Brain Res 1993; 625:177-79.

80. Lee S, Barbanel G, Rivier C. Systemic endotoxin increases steady-state gene expression of hypothalamic nitric oxide synthase: comparison with corticotropin-releasing factor and vasopressin gene transcripts. Brain Res 1995; 705:136-48.

81. Siaud P, Mekaouche M, Ixart G et al. A subpopulation of corticotropin-releasing hormone neurosecretory cells in the paraventricular nucleus of the hypothalamus also contain NADPH-diaphorase. Neurosci Lett 1994; 170:51-54.

82. Torres G, Lee S, Rivier C. Ontogeny of the rat hypothalamic nitric oxide synthase and colocalization with neuropeptides. Mol Cell Neurosci 1993; 4:155-63.

83. Villar MJ, Ceccatelli S, Bedecs K et al. Upregulation of nitric oxide synthase and galanin message- associated peptide in hypothalamic magnocellular neurons after hypophysectomy. Immunohistochemical and in situ hybridization studies. Brain Res 1994; 650:219-28.

84. Villar MJ, Ceccatelli S, Ronnqvist M et al. Nitric oxide synthase increases in hypothalamic magnocellular neurones after salt loading in the rat. An immunohistochemical and in situ hybridization study. Brain Res 1994; 644:273-81.

85. Vincent SR, Kimura H. Histochemical mapping of the nitric oxide synthase in the rat brain. Neuroscience 1992; 46:755-84.

86. Bredt DS, Glatt GE, Hwang PM et al. Nitric oxide synthase protein and mRNA are discrely localized in neuronal populations of the mammalian CNS together with NADPH diaphorase. Neuron 1991; 7:615-24.

87. Ceccatelli S, Hulting AL, Zhang X et al. Nitric oxide synthase in the rat anterior pituitary gland and the role of nitric oxide in regulation of luteinizing hormone secretion. Proc Natl Acad Sci USA 1993; 90:11292-96.

88. Ceccatelli S, Lundberg JM, Fahrenkrug J et al. Evidence for the involve-

ment of nitric oxide in the regulation of hypothalamic portal blood flow. Neuroscience 1992; 51:769-72.

89. Brunett AL, Ricker DD, Chamness SL et al. Localization of nitric oxide synthase in the reproductive organs of the male rat. Biol Reprod 1995; 52:1-7.

90. Iwai N, Hanai K, Tooyama I et al. Regulation of neuronal nitric oxide synthase in rat adrenal medulla. Hypertension 1995; 25:431-36.

91. Afework M, Ralevic V, Burnstock G. The intra-adrenal distribution of intrinsic and extrinsic nitrergic nerve fibres in the rat. Neurosci Lett 1995; 190:109-12.

92. Rettori V, Belova N, Dees WL et al. Role of nitric oxide in the control of luteinizing hormone-releasing hormone release in vivo and in vitro. Proc Natl Acad Sci USA 1993; 90:10130-34.

93. Bonavera JJ, Sahu A, Kalra PS et al. Evidence that nitric oxide may mediate the ovarian steroid-induced luteinizing hormone surge: involvement of excitatory amino acids. Endocrinology 1993; 133:2481-87.

94. Rettori V, Kamat A, McCann SM. Nitric oxide mediates the stimulation of luteinizing-hormone releasing hormone release induced by glutamic acid in vitro. Brain Res Bull 1994; 33:501-03.

95. Seilicovich A, Duvilanski BH, Pisera D et al. Nitric oxide inhibits hypothalamic luteinizing hormone releasing hormone relase by releasing γ-aminobutyric acid. Proc Natl Acad Sci USA 1995; 92:3421-24.

96. Duvilanski BH, Zambruno C, Seilicovich A et al. Role of nitric oxide in control of prolactin release by the adenohypophysis. Proc Natl Acad Sci USA 1995; 92:170-74.

97. Bonavera JJ, Sahu A, Kalra PS et al. Evidence in support of nitric oxide (NO) involvement in the cyclic release of prolactin and LH surges. Brain Res 1994; 660:175-79.

98. Rivier C, Shen GH. In the rat, endogenous nitric oxide modulates the response of the hypothalamo-pituitary-adrenal axis to interleukin-1β, vasopressin, and oxytocin. J Neurosci 1994; 14:1985-93.

99. Rivier C. Endogenous nitric oxide paricipates in the activation of the hypothalamic-pituitary-adrenal axis by noxious stimuli. Endocrine J 1994; 2:367-73.

100. Rivier C. Blockade of nitric oxide formation augments ACTH released by blood-borne interleukin-1β: role of vasopressin, prostaglandins and α-1 adrenergic receptors. Endocrinology 1995; 136:3597-603.

101. Costa A, Trainer P, Besser M et al. Nitric oxide modulates the release of corticotropin-releasing hormone from the rat hypothalamus in vitro. Brain Res 1993; 605:187-92.

102. Karanth S, Lyson K, McCann SM. Role of nitric oxide in interleukin 2-induced corticotropin-releasing factor release from incubated hypothalami. Proc Natl Acad Sci USA 1993; 90:3383-87.

103. Raber J, Koob GF, Bloom FE. Interleukin-2 (IL-2) induces corticotropin-releasing factor (CRF) release from the amygdala and involves a nitric oxide-mediated signaling; comparison with the hypothalamic response. J

Pharmacol Exp Ther 1995; 272:815-24.

104. Sandi C, Guaza C. Evidence for a role of nitric oxide in the corticotropin-releasing factor release induced by interleukin-1β. Eur J Pharmacol 1995; 274:17-23.

105. Coiro V, Volpi R, Chiodera P. Mediation by nitric oxide of TRH-, but not metoclopramide-stimulated TSH secretion in humans. Neuroreport 1995; 6:1174-76.

106. Kato M. Involvement of nitric oxide in growth hormone (GH)-releasing hormone-induced GH secretion in rat pituitary cells. Endocrinology 1992; 131:2133-38.

107. Rettori V, Belova N, Yu WH et al. Role of nitric oxide in control of growth hormone release in the rat. Neuroimmunomodulation 1994; 1:195-200.

108. Summy-Long JY, Bui V, Mantz S et al. Central inhibition of nitric oxide synthase preferentially augments release of oxytocin during dehydration. Neurosci Lett 1993; 152:190-93.

109. Yasin S, Costa A, Trainer P et al. Nitric oxide modulates the release of vasopressin from rat hypothalamic explants. Endocrinology 1993; 133:1466-69.

110. Yasin SA, Costa A, Hucks D et al. Interleukin-induced vasopressin release is inhibited by L-arginine. Ann NY Acad Sci 1993; 689:693-95.

111. Laskin JD, Heck DE, Laskin DL. Multifunctional role of nitric oxide in inflammation. Trends Endocrinol Metabol 1994; 5:377-82.

112. Moncada S, Palmer RMJ, Higgs EA. Nitric oxide: physiology, pathophysiology and pharmacology. Pharmacol Rev 1991; 43:109-42.

113. Lee S, Rivier C. Prenatal alcohol exposure alters the hypothalamic-pituitary-adrenal axis response of immature offspring to interleukin-1: is nitric oxide involved? Alcohol. Clin Exp Res 1994; 18:1242-47.

114. Brunetti L, Preziosi P, Ragazzoni E et al. Involvement of nitric oxide in basal and interleukin-1 beta-induced CRF and ACTH release in vitro. Life Sci 1993; 53:219-22.

115. Hashimoto K, Nishioka T, Tojo C et al. Nitric oxide plays no role in ACTH release induced by interleukin-1β, corticotropin-releasing hormone, arginine vasopressin and phorbol myristate acetate in rat pituitary cell cultures Endocrine J 1995; 42:435-39.

116. Sun Y, Rotenburg MO, Maines MD. Developmental expression of heme oxygenase isozymes in rat brain. J Biol Chem 1990; 265:8212-17.

117. Verma A, Hirsch DJ, Glatt CE et al. Carbon monoxide: a putative neural messenger. Science 1993; 259:381-84.

118. Stevens CF, Wang Y. Reversal of long-term potentiation by inhibitors of heme oxygenase. Nature 1993; 364:147-49.

119. Maines MD. Carbon monoxide: an emerging regulator of cGMP in the brain. Mol Cell Neurosci 1993; 4:389-97.

120. Pozzoli G, Mancuso C, Mirtella A et al. Carbon monoxide as a novel neuroendocrine modulator: inhibition of stimulated corticotropin-releasing hormone release from acute rat hypothalamic explants. Endocrinology

1994; 135:2314-17.

121. Parkes D, Kasckow J, Vale W. Carbon monoxide modulates secretion of corticotropin-releasing factor from rat hypothalamic cell cultures Brain Res 1994; 646:315-18.

122. Larmar CA, Mahesh VB, Brann DW. Regulation of gonadotropin-releasing hormone (GnRH) secretion by heme molecules: a regulatory role for carbon monoxide? Endocrinology 1996; 137:790-93.

CHAPTER 7

CYTOKINE ACTIONS ON BEHAVIOR

Robert Dantzer, Rose-Marie Bluthé, Arnaud Aubert, Glyn Goodall, Jean-Luc Bret-Dibat, Stephen Kent, Emmanuelle Goujon, Sophie Layé, Patricia Parnet and Keith W. Kelley

Sickness behavior refers to a coordinated set of behavioral changes that develop in sick individuals during the course of an infection. These changes are due to the effects of proinflammatory cytokines such as interleukin-1 (IL-1) and tumor necrosis factor alpha (TNF-α) on brain cellular targets and represent the expression of a well organized central motivational state. Based on the results of pharmacological and biochemical experiments, it is now apparent that sickness behavior is mediated by cytokines which are temporarily expressed in the brain in response to peripheral cytokines. Centrally released cytokines act on brain receptors which are identical to those characterized on immune cells. Primary afferent nerves represent the main communication pathway between peripheral and central cytokines. The sickness inducing effects of cytokines are downregulated by a number of endogenous neuropeptides and hormones, including vasopressin and glucocorticoids.

INTRODUCTION

Non specific symptoms of infection and inflammation include fever and profound physiological and behavioral changes. Sick individuals experience weakness, malaise, listlessness and inability to concentrate. They become depressed and lethargic, show little interest in their surroundings and stop eating and drinking. This constellation of nonspecific symptoms is collectively refered to as "sickness behavior." Due to their commonality, sickness symptoms are frequently ignored by physicians. They are considered as an uncomfortable, but rather banal, component of the pathogen-induced debilitative process.

A few years ago, Benjamin L. Hart, a veterinarian at the University of California at Davis, proposed that the behavioral symptoms of

Cytokines in the Nervous System, edited by Nancy J. Rothwell. © 1996 R.G. Landes Company.

sickness represent, together with the fever response, a highly organized strategy of the organism to fight infection.[1] This view has been built on the already recognized role of fever in the host response to pathogens. In physiological terms, fever corresponds to a new homeostatic state which is characterized by a raised set-point of body temperature regulation. A feverish individual feels cold at usual thermoneutral environments. Therefore it not only seeks warmer temperatures but also enhances heat production (increased thermogenesis) and reduces heat loss (decreased thermolysis). The higher body temperature that is achieved during fever stimulates proliferation of immune cells and is unfavorable for the growth of many bacterial and viral pathogens. In addition, the reduction of zinc and iron plasma levels that occurs during fever decreases the availablity of these vital elements for growth and multiplication of microorganisms. The adaptive nature of the fever response is apparent from studies showing that organisms infected with a bacteria or virus and unable to mount an appropriate fever response because they are kept in a cold environment or treated with an antipyretic drug have a lower survival rate than organisms which develop a normal fever.[2]

The amount of energy that is required to increase body temperature during the febrile process is quite high since, in human beings, metabolic rate needs to be increased by 13% for a rise of 1°C in body temperature. Because of the high metabolic cost of fever, there is little room for activities other than those favoring heat production (e.g., shivering) and minimizing thermal losses (e.g., rest, curl up posture, piloerection).

In recent years, evidence has rapidly accumulated to demonstrate that the necessary synchrony between metabolic, physiological and behavioral components of the systemic response to infection is dependent on the same molecular signals as those that are already responsible for the local immune reaction. These signals are proinflammatory cytokines such as interleukin-1 (IL-1), interleukin-6 (IL-6), tumor necrosis factor α (TNF-α) and interferons (IFNs), which are released by activated monocytes and macrophages during the course of an infection. IL-1 actually refers to two molecules, IL-1α and IL-1β, which have similar agonist activity on IL-1 receptors and are the product of separate genes. IL-1α is mainly secreted in a membrane-bound form whereas IL-1β is secreted in a soluble form.

The focus of the present chapter is exclusively on the effects of cytokines on behavior. We will show that the behavioral symptoms that develop during the host response to infection are mediated by the central actions of proinflammatory cytokines which are transiently expressed in the brain, in response to peripherally released cytokines.

CYTOKINES INDUCE SICKNESS BEHAVIOR

Studies of the biological activity of proinflammatory cytokines in vivo have been made possible by the availability of recombinant technology. Administration of recombinant human cytokines represents the most usual way of studying the specific biological activity of each of the proinflammatory cytokines. Although human recombinant IL-1 and IL-6 are fully active in laboratory animals, the same does not apply to TNF-α and IFN-α. Recombinant human TNF-α behaves as a partial agonist in the mouse system since it only binds to one of the two distinct TNF receptors, the 55 kDa TNF receptor. Recombinant human IFN-α is a poor agonist of the IFN-α receptor in the rat and mouse systems and homologous IFN-α should be used instead. Although most recombinant mouse cytokines are easily available, the same does not apply to rat cytokines. Administration of lipopolysaccharide (LPS), the active fragment of endotoxin from gram-negative bacteria, has the advantage of inducing the release of endogenous proinflammatory cytokines, of which the individual role can then be studied, using appropriate pharmacological tools such as specific antagonists of cytokines. Antagonists of cytokines used for pharmacological studies include antibodies to cytokines, antibodies to cytokine receptors, soluble receptors, natural antagonists to cytokine receptors, e.g., the IL-1 receptor antagonist, IL-1ra, and various other molecules.

Genetically modified mice in which the gene for a specific cytokine or cytokine receptor is either overexpressed or deleted (or "knocked-out") are an invaluable tool for the in vivo investigation of biological functions of cytokines. However, the redundancy of the cytokine network complicates the interpretation of the results obtained with these animal models.

The first evidence in favor of a possible role of cytokines in sickness behavior came from clinical trials with purified or recombinant cytokines in the treatment of intractable viral diseases and cancer.[3] Patients injected with these molecules were observed to develop not only flu-like symptoms but also, on repetition of injections, acute psychotic episodes characterized by depression or excitation. Fortunately, these symptoms spontaneously regressed on cessation of treatment. Toxicological studies carried out in laboratory animals confirmed the neurotropic activity of recombinant cytokines such as IL-1β and TNF-α. Animals injected acutely or chronically with these molecules usually appear lethargic, anorexic and withdrawn from their environment. The same effect is observed when animals are injected with non lethal doses of LPS.

More objective studies of the sickness inducing properties of cytokines have been based on the conditioned taste aversion (CTA) paradigm and the observation of profound changes in locomotor activity, social activities and food intake. The conditioned taste aversion paradigm is based on the association that animals form between the taste of the food or drink they have ingested and a subsequent episode of illness. In a typical experiment, rats are trained to drink their daily allocation of water during a 30 min presentation of a water bottle. On the day of conditioning, they are presented with a solution of saccharin instead of water and they are subsequently injected with a toxic agent. After recovery, they are presented with the saccharin solution that was paired with the poisoning episode, either alone or concurrently with water. Conditioned animals refrain from drinking the saccharin solution and the amount of saccharin drunk is an indirect measure of the intensity of the sickness previously experienced. In the CTA paradigm, systemic injection of rats with LPS after presentation of the new taste solution resulted in a decreased intake of this solution when it was presented later. Intraperitoneal (ip) or intracerebroventricular (icv) injections of IL-1β had the same effect.[4,5] Conditioned taste aversions have also been established to TNF-α. However, treatment with IFN-α failed to induce conditioned taste aversion.[6] The reasons for these differences between cytokines are still unclear.

Social activities can be investigated in many different ways. In rodents, olfactory exploration of conspecifics is an important component of social behavior since it allows the social partners to recognize their identity. Social exploration is followed by specific behavioral patterns such as courtship or aggression, according to the status of the conspecific. When juvenile conspecifics are presented to adult animals which are housed individually, exploratory behavior of this social stimulus by adults lasts for a few minutes and wanes off when the juvenile has become familiar. However, exploratory behavior is normally expressed again at a high level when unfamiliar juveniles are subsequently presented. This very robust behavior offers a simple way of testing the interference of sickness with social activities. Systemic administration of LPS, IL-1β and TNF-α to adult laboratory animals, rats or mice, decreases the duration of social exploration.[7-9] These effects take time to appear since they develop 2-4 hours after injection. They last for 4-6 hours and recovery is usually complete by 24 hours. A typical dose-response curve is shown in Figure 7.1.

Systemic administrations of IL-1 and TNF-α consistently suppress feeding and drinking. This effect has been observed using various measurements of food and water intake and ad libitum as well as deprived conditions.[10,11] Time course of the effects of cytokines on feeding behavior can be assessed by quantifying disruption of operant responding in animals trained to press a lever for food in a Skinner box. This technique was initially developed to objectively quantify sickness induced by removal of morphine or administration of the opiate receptor antagonist naloxone to morphine-dependent rats. Decreases in operant responding were found to be much more sensitive and specific than the physical symptoms which are often used to assess withdrawal (e.g., body weight and wet dog shakes).[12] Administration of LPS or recombinant cytokines disrupts food motivated behavior in a dose-dependent manner[10,13] (Fig. 7.2). A direct comparison of the effects of administration of IL-1β on social exploration and on food motivated behavior in rats revealed that the time course of action is

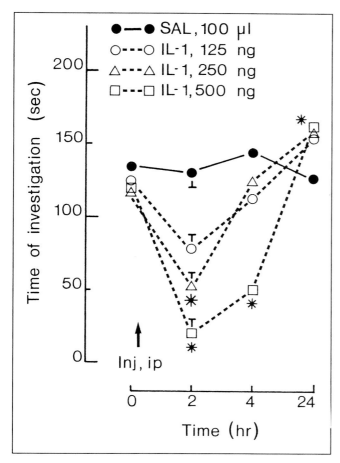

Fig. 7.1. Effects of intraperitoneal (ip) administration of IL-1β on social exploration in mice (mean + SEM, n = 5–6). Individually housed male mice were presented with a juvenile conspecific at time 0 and the duration of olfactory investigation of this social stimulus was assessed during a 4 min test. They were then injected with IL-1β and presented again with another juvenile at different time intervals. * p < 0.05 compared with value at time 0.

different according to the behavioral end-point. Decreases in food intake typically develop within one hour after injection whereas disruption of social exploration takes longer to appear.

An important component of sickness behavior is the increased somnolence that is observed during infectious episodes. The probability that this phenomenon is caused by the central effects of cytokines was first suggested by the observation that the endogenous molecules which are responsible for the increased sleepiness induced by sleep deprivation in experimental animals are bacterial cell wall products known as muramyl peptides.[14] Intravenous injection of IL-1β or TNF-α to rabbits induces dose-related increases in the amount of time spent in slow-wave sleep.[15] In rats, the effects of IL-1β on the architecture of sleep is more complex and depends on the circadian phase and the dose. For instance, low doses of IL-1β increased slow-wave sleep during the light period whereas higher doses increased wakefulness.[16]

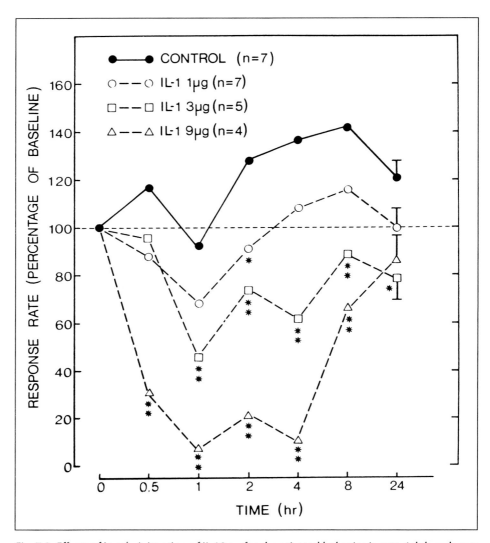

Fig. 7.2. Effects of ip administration of IL-1β on food-motivated behavior in rats. Adult male rats were deprived to 85% of their free-feeding body weight before being trained to press a bar for a 45 mg food pellet in a Skinner box on a fixed ratio 10 schedule (i.e., one food pellet for every 10 presses). Each rat served as its own control. Each test session lasted 5 min. Animals were injected with saline or different doses of IL-1β immediately after the first session (time 0). Data are expressed as percentage of preinjection response rate. Vertical bars represent SEM (*p < 0.05; ** p < 0.01 compared to respective control values) (From Bluthé et al, 1989).[13]

In view of the potent somnogenic effects of cytokines, it could be argued that the behaviorally depressing effects of cytokines just reflect the occurrence of sleeping episodes. However, this does not appear to be the case since LPS-treated rats which displayed a decreased response rate in a conditioning procedure in which the presentation of food was contingent on the intrusion of an operant lever in the cage whether the animal pressed the lever or not, still ate the presented food.[17] In the same manner, observation of IL-1β-treated rats shows that when they are exposed to a juvenile, they still respond to the juvenile when it comes in contact but do not follow it.

One of the cardinal symptoms of inflammation is pain. Because of the key role played by cytokines in inflammation, their effects on pain sensitivity has been assessed using different experimental paradigms. Hyperalgesia in response to IL-1β has been observed in four different model systems: the rabbit isolated ear perfusion model,[18] the paw pressure test in rats,[19] the tail-flick test and the formalin test.[20] In the rabbit isolated ear perfusion model, IL-1 enhanced the blood pressure effects induced by acetylcholine, which is typical of proalgic compounds. Intraplantar injection of IL-1 also resulted in an increased sensitivity to paw pressure.[21] This effect was not restricted to the site of injection since it was also observed in the contralateral paw. A similar hyperalgesic effect was also observed in the tail-flick test and the formalin test. Rats injected with ip LPS or IL-1β displayed prolonged hyperalgesia which developed within 5-10 min following injection and lasted for at least 1 hr. Although all these results are strongly suggestive of an enhancing influence of cytokines on pain sensitivity, analgesia following cytokine injection has also been observed using the hot plate test and the phenylquinone writhing test. A possible explanation for these contradictory effects is the difference in time of testing after injection rather than the technique used to assess pain sensitivity. In a systematic study of the time course of the modulatory influence of IL-1β on pain sensitivity, using the tail-flick test, IL-1β was found to increase pain sensitivity during the first hour after injection. This early hyperalgic response was followed by hypoalgesia later on.

MOTIVATIONAL ASPECTS OF SICKNESS BEHAVIOR

Sickness behavior is usually viewed by physicians as the result of debilitation and physical weakness which inevitably occur in an organism whose every resource is engaged in a defensive process against pathogens. An alternative hypothesis is that sickness behavior is the expression of a highly organized strategy that is critical to the survival of the organism. If this is the case, then it follows that sick individuals should be able to reorganize their behavior depending on its consequences and the internal and external constraint to which they are exposed. This flexibility is characteristic of what psychologists call a motivation. A motivation can be defined as a central state which reorganizes perception and action. A typical motivational state is fear. In order to escape a potential threat, a fearful individual must be attentive to everything which is occurring in his environment. At the same time, he must be ready to engage in the most appropriate defensive behavioral pattern which he has available in his behavioral repertoire. In other terms, a motivational state does not trigger an unflexible behavioral pattern. It enables one to uncouple perception from action and therefore to select the appropriate strategy depending on the eliciting situation.[22]

The first evidence that sickness behavior is the expression of a motivational state rather than the consequence of weakness was provided by Neal Miller, a psychologist working at Rockfeller University.[23] While studying the mechanisms of thirst, he was struck by the observation that thirsty rats injected with endotoxin stopped bar pressing for water but, when given water, drank it although to a lesser extent than normally. This effect was not specific to thirst since the endotoxin treatment also reduced bar pressing for food and even blocked responding in rats trained to press a bar for the rewarding effects of electrical stimulation in the lateral hypothalamus (self-stimulation). Interestingly enough, when rats were trained to turn off an aversive electrical stimulation in this brain area, endotoxin also reduced the rate of responding, but to a lesser extent than bar pressing for a rewarding brain stimulation. More to the point, however, was the observation that rats which were placed in a rotating drum could stop for brief periods by pressing a lever which increased their response rate in response to endotoxin treatment.

The mere fact that endotoxin treatment can decrease or increase behavioral output depending on its consequences gives strong support to the motivational interpretation of the behavioral effects of such a treatment.

To further support this interpretation, rats were trained to work for food on a progressive ratio schedule. In the schedule which was chosen for this particular experiment, they had to press the lever once to get the first pellet, twice to get the second pellet, four times to get the third pellet, eight times for the fourth pellet, and so on. This means that food became more and more difficult to obtain with time. A normal animal stops pressing the lever when the ratio becomes too high in comparison to its motivation for food. If food motivation is decreased, like after a free meal, the achieved ratio gets smaller whereas if food motivation increases, like after a longer duration of food deprivation, the achieved ratio gets larger. When well trained rats had to work for food in a cold environment (4-5°C), some of them were observed to increase their response rate whereas the other animals responded less. These divergent changes correspond to two different thermoregulatory strategies, consisting of increasing energy intake to produce more heat in the first case, and decreasing energy expenditure and minimizing thermal losses by stopping lever pressing and adopting a curled posture in the second case. The important result is that when rats were injected with a small dose of IL-1β directly into the lateral ventricle of the brain, they responded to this immune signal in exactly the same way as when they were exposed to cold, i.e., those individuals which enhanced their response rate in the cold enhanced their response rate in reponse to IL-1β and vice-versa.

Another important characteristic of a motivational state is that it competes with other motivational states for behavioral output. As a typical example, it is somewhat difficult to search for food, and at the same time court a sexual partner since the behavioral patterns of foraging and courtship are not compatible with each other. The normal expression of behavior therefore requires a hierarchical structure of motivational states which is continuously updated according to the urgencies that occur in the internal and external milieux. When an infection occurs, the sick individual is at a life or death juncture and its physiology and behavior must be altered so as to overcome the disease. However, this is a relatively long-term process which needs to make room for more urgent needs when necessary. It is easy to imagine that if a sick person lying in his bed hears a fire alarm ringing in his house and sees flames and smoke coming out of the basement, he should be able to momentarily overcome his sickness behavior to escape danger. In motivational terms, fear competes with sickness, and fear motivated behavior takes precedence over sickness behavior. An example of this competition between fear and sickness is provided by the observation that the depressing effects of IL-1β on behavior of mice are more pronounced when experimental animals are tested in the safe surroundings of their home cage than when they are placed into a new environment.

Another example of the motivational aspects of sickness behavior is the effects of cytokines on maternal behavior. If fitness is the key issue, it is evident that dams should care for their youngs despite sickness. In motivational terms, the components of maternal behavior that are crucial for the survival of the progeny should be more resilient, i.e., less sensitive to the depressing effects of pyrogens, than those behavioral patterns that are less important. In accordance with this prediction, administration of LPS to lactating pluriparous mice did not disrupt pup retrieval but impaired nest-building. However, when the dams and their litters were exposed to 4°C, to increase the fitness value of nest building, this activity was less disrupted by LPS than when dams were tested at 20°C (Aubert et al, in preparation).

From an adaptive point of view, the anorexic effect of cytokines is difficult to reconcile with their pyrogenic activity. The decrease in food intake that acompanies fever appears to be inconsistent with the enhanced energy requirement of thermogenesis. To resolve this paradox, it has been proposed that cytokine-induced anorexia spares energy required for foraging and prevents a weakened organism to run into the risk of being exposed to a predator during the search for food. If this is the case, cytokines should be more effective to suppress the foraging than the consummatory components of food intake. There is some evidence that cytokines have such a differential effect since, as previously mentioned, LPS- or IL-1-treated animals stop pressing a lever for food but still eat the food pellets which are delivered independently of their behavior. However, a more direct assessment of the effects of proinflammatory cytokines on the foraging and consummatory components of feeding behavior is still lacking.

Food intake is altered by cytokines not only in a quantitative but also in a qualitative way. It has been amply demonstrated that when rats are given the opportunity to select components of their diet, their selection pattern reflects the organism's nutritional and energetic requirements. To determine whether this selection pattern is altered during sickness, rats were submitted to a dietary self selection protocol in which they had free access to carbohydrate, protein and fat diets for 4 hours a day.[24] After a 10-day habituation to this regimen, they were injected with LPS or IL-1β. Under the effect of this treatment, they decreased their total food intake but reorganized their self selection pattern so as to ingest relatively more carbohydrate and less protein, whereas fat intake remained unchanged. This change in macronutrient intake contrasts with the increased fat intake that occurs in rats exposed to cold. Although eating fat would be a better way for feverish animals to account for their increased energy requirements, it would not be of much use since cytokines have profound metabolic effects resulting in increased lipolysis and hypertriglyceridemia. Under these conditions, an increased intake of fat would actually be counterproductive since it would further enhance hyperlipidemia without positively contributing to lipid metabolism.

MECHANISMS OF THE BEHAVIORAL EFFECTS OF CYTOKINES

Role of Endogenous Cytokines

Proinflammatory cytokines act in a cascade fashion, each cytokine being able to induce its synthesis and the synthesis of other cytokines. Because of the pleiotropism and redundancy of the cytokine network, it is important to determine which cytokine contributes to sickness behavior and whether different cytokines mediate different components of sickness behavior. In order to answer these questions, it is not sufficient to test the effect of different cytokines on various behavioral patterns. It is also necessary to block the activity of the cytokine under investigation. As pointed out earlier, administration of LPS induces the synthesis and release of most proinflammatory cytokines. In rats injected with LPS, depression of social exploration was blocked by pretreatment with the specific antagonist of IL-1 receptors, IL-1Ra[7] (Fig. 7.3). However, the same pretreatment had no effect on the LPS-induced decrease in food motivation.[10] These results indicate that IL-1 is the main mediator of the social component of sickness behavior but that other cytokines mediate the alterations in food intake that occur in sick individuals. Concerning the possible involvement of IL-1 in the effects of other cytokines, the observation that the injection of TNF-α mimicked the depressing effect of IL-1β on social exploration and synergized with IL-1β to induce sickness behavior led to the investigation of the effect of IL-1ra on the behavioral effect of TNF-α. Pretreatment with IL-1ra abrogated the behavioral effect of TNF-α, which can be interpreted to suggest that TNF-α acts on social exploration by inducing the synthesis and release of IL-1.[9] However, since recombinant human TNF-α was used in this experiment, an alternative explanation is that activation of the p55 receptor is not sufficient to alter behavior and requires cooperation with IL-1 receptors.

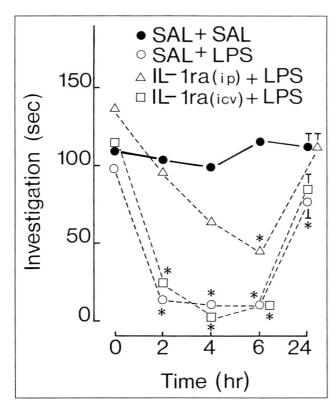

Fig. 7.3. Effects of IL-1ra on LPS-induced depression of social exploration in rats. Rats were injected with either saline or IL-1ra followed by saline or LPS. Four experimental groups were used: saline (ip) + saline (ip) (n = 5); saline (ip) + LPS (250 μg/kg ip) (n = 6); IL-1ra (3 mg/rat ip) + LPS (250 μg/kg ip) (n = 5); IL-1ra (60 μg/rat icv) + LPS (250 μg/kg ip) (n = 7). Injections were administered immediately after the first session, in which their interest toward a juvenile conspecific was measured, and animals were tested again after 2, 4, 6 and 24 h. SEM is represented only for the last point. *p < 0.05 compared with values at time 0 (From Bluthé et al, 1992b).[7]

CENTRAL VERSUS PERIPHERAL SITES OF ACTION OF CYTOKINES ON BEHAVIOR

Since cytokines are released locally in the body during the course of an infection or inflammation, the question of how these molecules act on the brain has been the object of much debate during the last decade. Cytokines are relatively big proteins (about 150 amino acids for IL-1β) which are hydrophilic and therefore cannot cross the blood-brain barrier. For this reason, they are believed to act on those brain sites which lack a blood-brain barrier and are known as circumventricular organs because of their spatial location close to the brain ventricles. There, cytokines would trigger the synthesis and release of prostaglandins of the E series which would freely diffuse to the target neuronal cells. This mode of action has been proposed to account for both the pyrogenic and corticotropic activity of cytokines.[25,26]

The problem with this hypothesis is that it does not account for the observation that cytokines are present with their receptors in the brain and their local expression is modulated by peripheral immune stimuli.[27,28] There are several lines of evidence to support the hypothesis that central cytokines are responsible for the behavioral effects of peripherally released cytokines. Indirect evidence comes from the observation that much lower doses of IL-1β or TNF-α need to be injected directly into the lateral ventricle of the brain or specific brain structures to depress feeding behavior and social exploration than the doses which are required when the same cytokine is injected at the periphery. The ratio is usually 1 to 100 or 1000.[3] More direct evidence for a central site of action of proinflammatory cytokines is the demonstration of the blockade of the behavioral effects of peripherally injected IL-1β by a central injection of the specific antagonist of IL-1 receptors, IL-1ra.[29] In this experiment, rats were pretreated icv with a dose of IL-1ra sufficient to block the depressing effects of icv injected IL-1β on social exploration and food-motivated behavior, but which had no antagonistic effect when injected at the periphery. This pretreatment completely abrogated the reduction of social exploration induced by ip IL-1β and attenuated the decrease in response rate that occurs in IL-1β-treated rats trained to get their food by pressing a lever in a Skinner box (Fig. 7.4). These results indicate that IL-1 administered at the periphery acts in the brain via classical IL-1 receptors. However, they do not allow us to determine where the centrally acting IL-1 comes from, i.e., whether it is imported from the periphery or produced locally in the brain. The demonstration that peripheral administration of LPS or IL-1 induces the expression of cytokines in the brain at the mRNA and protein levels[27,30,31] is strongly suggestive of the second possibility.

PERIPHERAL CYTOKINES ACTIVATE THE SYNTHESIS AND RELEASE OF CENTRAL CYTOKINES VIA NEURAL AFFERENT PATHWAYS

Based on the previous findings, it has been proposed that cytokines released at the periphery activate primary sensory neurons and

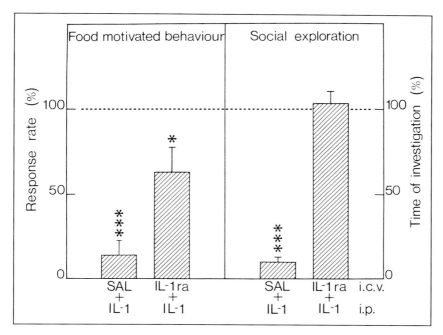

Fig. 7.4. Differential effects of blockade of central IL-1 receptors on the effects of IL-1β on food-motivated behavior and social exploration in rats. IL-1ra or saline was injected icv into rats trained to press a lever for food on a fixed ratio 10 schedule (24 µg of IL-1ra /rat) or into rats presented with a juvenile conspecific (4 µg of IL-1ra/rat). This injection was followed by ip IL-1β (4 µg/rat) or saline. Injections were given immediately after the first test session and animals were tested again 1 h (food-motivated behavior) or 2 h (social exploration) later. The figure represents percentage of variation with regard to baseline values (* p < 0.05, *** p < 0.001). Note that pretreatment with IL-1ra blocked the effects of IL-1β on social exploration but only partially attenuated the effects of this cytokine on food-motivated behavior (From Kent et al, 1992a).[29]

that this neuronal activation conveys the peripheral immune message to the brain where it results in the synthesis and release of a functional pool of cytokines.[3] In accordance with this hypothesis, intraperitoneal administration of LPS has been found to increase the levels of sensory neuropeptides (substance P, neurokinin A and calcitonin-gene-related peptide) in the spinal cord[32] and induce the expression of the cellular immediate-early gene *c-fos* in various areas of the brain.[33] *c-fos* has been validated as an inducible and widely applicable marker for neural systems activated by a variety of extracellular stimuli. The protein product Fos of this gene interacts with nuclear proteins to act as a transcription factor. In response to ip injection of LPS, Fos immunoreactive neurons have been identified in the primary projection area of the afferent branches of the vagus nerve, represented by the nucleus tractus solitarius, and secondary projection areas which include, among others, the parabrachial nucleus, the paraventricular nucleus and the supraoptic nucleus of the hypothalamus. Furthermore, transection of the vagus nerve at the subdiaphragmatic level, which eliminates afferent neurons orginating from the liver and the gastro-intestinal tract, abrogates LPS-induced Fos immunoreactivity in these brain areas.[33]

The functional nature of this vagal communication pathway between the immune system and the brain has been evidenced by the demonstration that section of the vagus nerve abrogates LPS-induced hyperalgesia[34] in rats as well as LPS- and IL-1β-induced decreases in social exploration and food-motivated behavior in rats and mice[35,36] (Fig. 7.5). The role of the vagus nerve does not appear to be limited to the behavioral effects of cytokines since fever and pituitary-adrenal activation in response to LPS are also blocked in vagotomized rats.[37,38] In accordance with such a wide range of effects of vagotomy on the neural activity of peripheral immune stimuli, this surgical procedure has been found to result in the abrogation of the induction of IL-1β in the brain at the mRNA and protein levels in response to peripherally injected LPS.[39]

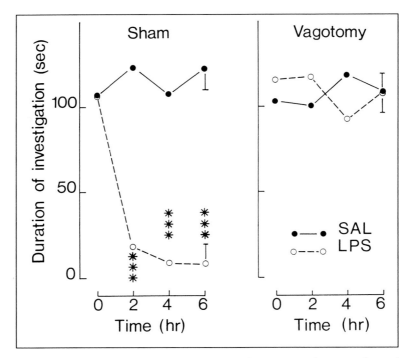

Fig. 7.5. Effects of vagotomy on LPS-induced decreases in duration of social exploration. LPS (1.25 mg/kg) or physiological saline were injected to sham-operated (left figure) and vagotomized (right figure) rats immediately after the first behavioral session which took place at time 0 and the same animals were tested again with different juveniles 2, 4 and 6 h after (n = 4 for each experimental group, except LPS-treated sham animals for which n = 3) (from Bluthé et al, 1994b).[35] Note that LPS significantly decreased social exploration only in sham operated rats (*** p < 0.001 compared to saline value).

The observation that a section of the vagus nerve blocks the effects of intraperitoneal administration of IL-1β and LPS is in agreement with the role of this nerve in many visceral reflexes. However, the vagus nerve is not classically considered as a pathway involved in the transmission of visceral sensitivity. This type of information reaches the brain via the splanchnic nerve and the spinal cord. Section of the splanchnic nerve did not attenuate the hyperalgic activity of LPS,[34] which can be interpreted to suggest that this nerve pathway plays no role in the transmission of immune signals to the brain.

The problem with vagotomy is that this surgical procedure, even when it is carried out below the diaphragm, is a relatively drastic procedure which might attenuate the effects of LPS and IL-1 because it alters the sensitivity of both peripheral and central targets to the effects of cytokines. Vagotomized animals need at least one week to recover from surgery and their food intake remains disturbed for several weeks because of the slower emptying of the stomach. The stomach is enlarged and fills up most of the abdominal cavity. In view of these changes, it could be argued that vagotomized animals are less sensitive to the effects of intraperitoneal administration of LPS and IL-1 simply because the sensitivity of their peripheral target organs is impaired. This does not appear to be the case, since the increased levels of IL-1β in macrophages and plasma in response to ip LPS was not attenuated by vagotomy.[35] Another possibility is that sensitivity of the central targets for the effects of cytokines produced in the brain is decreased because of the cellular reorganization that takes place in the primary and secondary projection areas of the vagus nerve after section of this cranial nerve. If it is the case, vagotomized rats should be less sensitive to the effects of centrally injected IL-1β. This was not the case since injection of IL-1β into the lateral ventricle of the brain decreased social exploration of juveniles by vagotomized rats to the same extent as the decrease which was observed in sham operated rats, whereas intraperitoneal administration of IL-1β had no effect.[40] The observation that vagotomy blocks only the behavioral effects of IL-1 when this cytokine is injected intraperitoneally but not when it is administered via the subcutaneous or the intravenous route[41] gives further support to the specificity of the role of the vagus nerve in transmitting the immune information from the abdominal cavity to the brain.

The findings from vagotomy experiments are important because they indicate that the brain is able to sense that cytokines have been released at the periphery during the course of an infection or inflammation and to respond to this stimulus by a local synthesis of cytokines. The mechanisms that are responsible for the transformation of the immune message in a neural message at the periphery and the transduction of this neural message back into an immune message in the central nervous system still need to be worked out, but they provide a fascinating example of communication between the brain and the periphery (Fig. 7.6).[42]

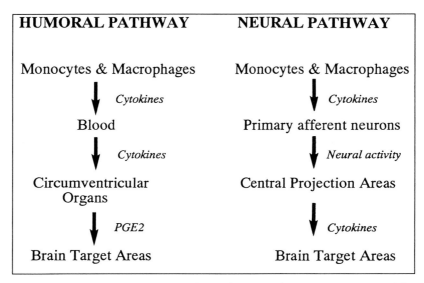

HUMORAL PATHWAY   **NEURAL PATHWAY**

Monocytes & Macrophages   Monocytes & Macrophages

↓ *Cytokines*   ↓ *Cytokines*

Blood   Primary afferent neurons

↓ *Cytokines*   ↓ *Neural activity*

Circumventricular   Central Projection Areas
Organs

↓ *PGE2*   ↓ *Cytokines*

Brain Target Areas   Brain Target Areas

Fig. 7.6. Possible communication pathways between the immune system and the brain. In the humoral interpretation of the effects of cytokines on brain functions, cytokines are released into the general circulation by activated monocytes and macrophages. They reach circumventricular organs where they induce the synthesis and release of prostaglandins which freely diffuse into the brain parenchyma to reach cell targets mediating the various neural effects of cytokines. In the neural interpretation of the effects of cytokines on brain functions, locally released cytokines activate peripheral afferent nerves. The subsequent changes in firing rate are transmitted to the brain where they induce the synthesis and release of cytokines by resident macrophages and microglial cells. The cell targets for the cytokines which are produced in the brain are still unknown.

ROLE OF CORTICOTROPHIN RELEASING FACTOR (CRF) IN THE BEHAVIORAL EFFECTS OF CYTOKINES

Cytokines have potent activating effects on the pituitary-adrenal system which are mediated via the release of CRF. The possible involvement of this hypothalamic neuropeptide in the neural effects of IL-1 has been assessed by icv administration of CRF antiserum or a peptide known as α-helical CRF(9-41) and behaving as an antagonist of CRF receptors (αhCRF(9-41)).

Immunoneutralization of endogenous CRF in the brain attenuated the anorexic effects of IL-1β.[43] The same treatment blocked the reduction of immobility induced by icv IL-1β in rats forced to swim in a confined space.[44] Central administration of αhCRF(9-41) was also able to prevent the IL-1β-induced reduction of exploratory behavior of mice placed in a multicompartment chamber.[45]

Although these findings can be interpreted to suggest that brain CRF mediates the behavioral effects of IL-1, there is evidence contradicting this conclusion. In particular, the decrease in food-motivated behavior which was induced in rats by ip injection of IL-1β was not altered by icv administration of either αhCRF(9-41) or CRF itself.[13] When IL-1β was injected centrally, αhCRF(9-41) did not alter the peak effect but facilitated the return to baseline.[40]

The exact factors that are responsible for these differences are still unknown. However, it is clear that the involvement of CRF in the behavioral effects of IL-1 is not a general phenomenon.

ROLE OF PROSTAGLANDINS IN THE BEHAVIORAL EFFECTS OF CYTOKINES

IL-1 and other proinflammatory cytokines are potent inducers of prostaglandin production. Administration of cyclooxygenase inhibitors such as indomethacin at doses that abolish synthesis of prostaglandins attenuated the pyrogenic and anorexic effects of IL-1. Pretreatment with indomethacin or piroxicam blocked the depressing effects of ip administration of IL-1β on food-motivated behavior in rats and social exploration in mice.[46] Whether these effects are mediated peripherally or centrally still needs to be elucidated. Central administration of ibuprofen failed to block the anorexia induced by centrally administered IL-1β although it attenuated the increase in body temperature caused by central IL-1β.[47] In contrast, ip injection of ibuprofen partially blocked the anorexic effects of icv IL-1β.

In rats continuously infused with recombinant murine IL-1α, piroxicam completely inhibited the stimulation of drinking behavior, but had no effect on the reduction in eating activity and locomotor activity induced by this cytokine.[48] Pretreatment with indomethacin had no effect on the depression of general activity and food intake induced in mice by peripheral injection of IFN-α.[49]

In the face of these contradictory results, there is clearly a need for further studies on the role of prostaglandins in the behavioral effects of cytokines.

ROLE OF NITRIC OXIDE IN THE BEHAVIORAL EFFECTS OF CYTOKINES

The sustained vasodilatation and hypotension induced by IL-1 and other proinflammatory cytokines are mediated by the local synthesis and release of nitric oxide (NO) via induction of the type II NO synthase in both endothelial and vascular smooth muscle cells. In addition to its potent vasodilatory activity, NO behaves as an effector molecule in immunological reactions, and as a neurotransmitter in the central and peripheral nervous system.

Administration of agents that block synthesis of NO from L-arginine attenuated the dramatic fall in blood pressure that occurs in septic shock or in response to exogenously administered cytokines. In the same manner, inhibition of activity of NO synthase attenuated somnogenic effects of cytokines. However, the same treatment had no effect on fever.

To test whether NO production is also involved in the behavioral effects of IL-1β, mice were pretreated with various doses of N-nitro-L-arginine methyl ester (NAME), a selective inhibitor of the brain and endothelial NO synthases.[8] Administration of a high dose of the antagonist potentiated the depressing effects of IL-1β on social exploration, whereas a lower dose had no effect. This potentiation was due to the inhibition of the synthesis of NO since it was attenuated by L-arginine but not by D-arginine (Fig. 7.7). Administration of the nitro-arginine derivative had no effect on the weight loss induced by IL-1β. Although these results indicate that NO production might have a protective role on the effects of IL-1 on brain functions, they need to be complemented by further studies assessing the effects of IL-1 on brain NO synthases.

OTHER MEDIATORS OF THE BEHAVIORAL EFFECTS OF CYTOKINES
Cholecytokinin (CCK) is a peptide which plays an important role in the regulation of food appetite and gastric emptying. CCK is released by the small intestine during feeding. It acts on CCK receptors in the gut, which stimulates neural structures in the brain that control feeding behavior via vagal efferent fibers, resulting in cessation of food consumption. CCK receptors are of two types, the type A receptor which is located primarily in the gut, and the type B receptor which is located mainly in the brain. The anorexic effects of CCK appear to be mediated via type A receptors.

Since the anorexic effects of IL-1 resemble those of CCK, the role of CCK in IL-1 induced anorexia has been investigated indirectly by measuring the effects of IL-1 on plasma levels of CCK and directly, by administering a CCK-A receptor antagonist (L364,718) before injection of IL-1α. Administration of IL-1α significantly increased plasma levels of CCK and decreased food intake and gastric emptying in rats. These last effects were partially blocked by pretreatment with L364,718.[50]

OPPOSITION OF THE BEHAVIORAL EFFECTS OF CYTOKINES BY CRYOGENS
At the same time that they affect brain functions, cytokines trigger neurotransmitter mechanisms which are part of a feedback loop regulating the neurotropic effects of cytokines and promoting recovery. Since these neurotransmitters have originally been identified on the basis of their ability to oppose the pyrogenic effects of cytokines, they have been named endogenous pyrogens or cryogens.[51]

The neuropeptide vasopressin (VP) is one of these substances. It is mainly known as a key factor in regulation of water metabolism. It is produced by magnocellular hypothalamic neurons and accumulates in the terminals of these neurons in the posterior pituitary. It is re-

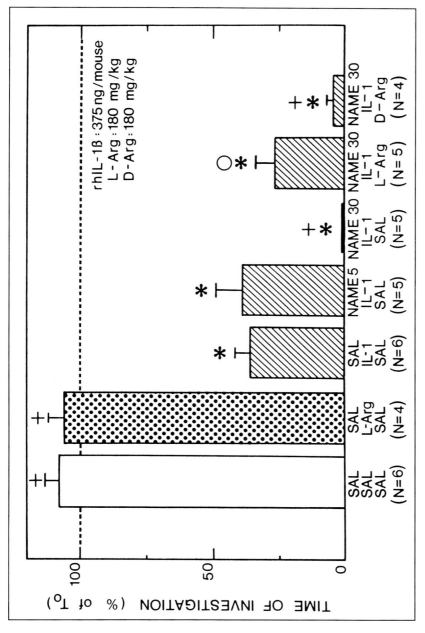

Fig. 7.7. Potentiation of IL-1β-induced changes in social exploration by pretreatment with NAME, an inhibitor of NO synthase, and the effects of L-arginine or D-arginine. Each column represents the mean variation measured 4 h after injection of IL-1β. NAME was injected at a dose of 5 or 30 mg/kg. Previous experiments had established that these doses of L-NAME had no effect on social exploration. The number of mice in each group is given in parentheses. * p < 0.05 with comparison to saline; + p < 0.05 with comparison to IL-1; O p < 0.05 with comparison to the (IL-1 plus NAME) group (From Bluthé et al, 1992c).[8]

leased in the general circulation and acts as an antidiuretic hormone in conditions of water deprivation and in response to fever. VP can also act as a neurotransmitter in the brain: it is present in particular in neurons of which the cell bodies are located in the bed nucleus of the stria terminalis (BNST) and the terminals project to the lateral septum. This vasopressinergic pathway is highly sensitive to circulating androgens in rodents. Castration leads to a dramatic reduction in the content of VP mRNA in the BNST neuronal cell bodies and to a reduction in immunoreactive VP in the terminal areas of the septum. This androgen-dependent pathway is also activated during fever.

Central administration of VP attenuated the depressing effects of centrally injected IL-1β on social exploration. Conversely, central injection of an antagonist of vasopressin receptors, which has no biological activity on its own but prevents endogenously released vasopressin to reach its receptors, sensitized rats to the behavioral effects of IL-1. These last results are important since they suggest that endogenous VP play a physiological modulatory role in the behavioral effects of IL-1. To determine whether this phenomenon is mediated by an androgen-dependent or independent pathway, castrated male rats were compared to intact male rats in their sensitivity to the modulatory role of the VP receptor antagonist. Castration by itself potentiated the depressing effects of IL-1β on social exploration. Central administration of VP was more effective in attenuating the behavioral effects of IL-1 in castrated than in intact male rats and, conversely, icv administration of the VP receptor antagonist was no longer active in potentiating the behavioral effects of IL-1 in castrated male rats lacking vasopressinergic innervation of the lateral septum.[52]

The mechanisms by which the brain vasopressinergic system is activated by cytokines and the way vasopressin interacts with the effect of cytokines on their target cells remain to be elucidated.

The potent activating effects of proinflammatory cytokines on pituitary-adrenal activity have already been mentioned. Glucocorticoids represent another class of key molecules in the regulation of sickness behavior. Adrenalectomy is acompanied by an increased sensitivity to the depressing effects of IL-1β and LPS on social exploration. This effect was mimicked by acute administration of the glucorticoid receptor antagonist RU-38486 to intact mice and it was abrogated by the implantation of a corticosterone pellet in adrenalectomized mice.[53]

Adrenalectomized animals implanted with a corticosterone pellet have constant levels of corticosterone but are unable to respond to administration of IL-1 by enhanced pituitary-adrenal activity. This phasic response to IL-1 appears to be important in regulating the behavioral effects of cytokines since the protection offered by a corticosterone pellet which ensured plasma levels of corticosterone intermediate between normal and stress levels was effective only against low doses but not high doses of IL-1β.

The effects of glucocorticoids on the behavioral activity of IL-1 and LPS are likely to be mediated by both peripheral and central mechanisms. Adrenalectomized mice were also more sensitive to the depressing effects of a central injection of IL-1β on social exploration and icv administration of the glucocorticoid receptor antagonist RU-38486 mimicked this effect.[54] The mechanism underlying these effects is represented by the downregulation action of glucocorticoids on the synthesis and release of cytokines by both peripheral immune cells and brain cells.[55]

PATHOPHYSIOLOGICAL IMPLICATIONS OF THE BEHAVIORAL EFFECTS OF CYTOKINES

The demonstration that the immune system is able to influence behavior and mental states has important implications on our understanding of the relationships between psychological factors and disease. In the case of cancer, for example, such psychological features as the feelings of hopelessness and helplessness that are commonly associated with the onset and progression of the disease might be secondary to the effects on the central nervous system of factors released from immune or tumor cells during the early stage of the tumoral process. The same possibility applies to the relationship between psychological factors and autoimmune diseases. The possible causal role of cytokines in the mental and behavioral symptoms that occur in various pathological conditions, has hardly been investigated, except in a few cases, such as infection and fever, cachexia, AIDS dementia complex, chronic fatigue syndrome and depression.

There is already evidence demonstrating that proinflammatory cytokines are responsible for the development of subjective and behavioral symptoms of sickness during infection with a bacterial or viral pathogen. For instance, patients treated with IFN-α showed fever, anorexia, fatigue, headache, myalgia and arthralgia. These symptoms culminated in lethargy and withdrawal from the surroundings. The same symptoms were observed in volunteers injected with low doses of LPS.

The possibility that the release of cytokines accounts for more subtle changes in cognition and performance has been assessed by Smith, a psychologist working at the MRC Common Cold Unit, in Brighton, UK. On the basis of earlier work showing that infection with upper respiratory viruses decreased the efficiency with which psychomotor tasks were performed, volunteers of both sexes were injected with IFN-α. Volunteers injected with the larger dose were significantly slower at responding in a reaction time task when they were uncertain when the target stimulus would appear. Simultaneously, they displayed hyperthermia and experienced feelings of illness. However, they were not impaired on a pursuit tracking task or syntactic reasoning task. These effects were similar to the alterations in performance observed in patients with influenza.[56] The possibility that proinflammatory cytokines have relatively specific effects on cognitive processes has been further investigated in animal models. IL-1β, but not IL-6, impaired spatial navigation learning in rats.[57] A similar deficit in spatial learning was observed in mice injected with IL-1β or infected with the pathogenic agent *Legionella pneumophilia*.[58] Interference of cytokines with formation of new memories has also been demonstrated in an autoshaping task in which rats learned to press a lever which was introduced into the cage before food delivery.[17] These effects of cytokines appear to be independent of their pyrogenic activity since they were observed whether body temperature increased or decreased in response to the treatment under study.

There has been much speculation on the possible pathogenic role of cytokines in the chronic fatigue syndrome (CFS). Always feeling tired is a common complaint in patients afflicted with a viral infection and represents the core symptom of the so-called postviral fatigue syndrome. CFS patients feel the same but in the absence of any persistent viral infection.[59] Their symptoms are real, pervasive and often incapacitating. The fact that a substantial proportion of these patients fulfill criteria for major depression and other psychiatric illness does not facilitate the classification of this disorder. Whatever the case, and in view of the similarities between the subjective effects of cytokines and the symptoms reported by CFS patients, many researchers have looked for possible hyperproduction of cytokines in this condition. Elevated plasma levels of IL-2 and IL-6 have been reported in one study of CFS patients, but this result has not been found in other studies. Such inconsistent results can be easily explained by technical problems associated with the detection of cytokines in biological fluids and the poor correlation between plasma levels of cytokines and local activity of these mediators. A better way of assessing peripheral cytokine function is to study the ability of peripheral blood mononuclear cells (PBMC) to produce cytokines when put in culture and stimulated with LPS. Using such a strategy, Chao et al[60] reported an hyperproduction of IL-6 and TNF-α and a lower production of the inhibitory cytokine TGFβ in CFS patients. However, it is important to note that abnormalities, if any, of the cytokine network, are not necessarily present at the periphery but might instead preferably affect cytokines that are expressed in the central nervous system. Although this possibility is much more difficult to put to test, it has inspired quite elaborate speculation.[62]

Besides their commonality in CFS patients, lack of energy and loss of interest are very frequent in depressed patients. These symptoms are actually incorporated in the basic description of depressive episodes. The 10th revision of the International Classification of Disease begins with the statement that "the subject suffers from a lowering of mood, reduction of energy, and decrease in activity. Capacity for en-

joyment, interest and concentration are impaired, and marked tired-
ness after even minimum effort is common." The possibility that activation of peripheral blood monocytes and T lymphocytes play a role
in the pathophysiology of major depression has been proposed by Michael
Maes, a Belgian psychiatrist.[62] In addition to the evidence pointing
out the profound effects of cytokines on behavior and the hypothalamic-pituitary-adrenal axis, this hypothesis is based on the observation of an increased production of cytokines by monocytes and T lymphocytes of depressed patients. For example, elevated levels of acute
phase proteins and increased concentrations of IL-6 and its soluble
receptor have been found in the plasma of subjects with major depression and there was a close relationship between IL-6 levels and acute
phase proteins. However, more research is still needed before a role of
immune products in the pathogenesis of depressive symptoms can be
accepted. The observed immune alterations appear to be a trait rather
than a state marker of depression since they persist even when depressive symptoms regress. In addition, the possible contribution of anti-
depressant treatment to the changes in immune functions observed in
depressed patients remains to be established.

The possibility of a role of cytokines in depression has also been
studied in animal models of depression. In the absence of any knowledge on the causal factors of depression, most animal models of depression are based on behavioral and pharmacological analogies. At the
behavioral level, the two main symptoms that are usually considered
include the deficit in escape/avoidance learning and the anhedonia, or
more precisely the diminished capacity to experience pleasure, which
are typically displayed by experimental animals exposed to uncontrol-
lable electric shocks. At the pharmacological level, chronic but not
acute treatment with antidepressant drugs blocks the development of
these symptoms. Two recent studies provide some evidence for a role
of cytokines in animal models of depression. Intracerebroventricular
administration of the interleukin-1 receptor antagonist (IL-1ra) which
blocks the access of endogenous IL-1 to its receptors, attenuated the
escape-avoidance deficit induced by inescapable electric shock in rats.[63]
In a different series of experiments, systemic administration of LPS
was found to induce anhedonia, as evidenced by an attenuated prefer-
ence for a saccharin solution. This effect was antagonized by chronic
but not acute treatment with the anti-depressant drug imipramine.[64]
Even if they are certainly more related to the role of endogenous cytokines in response to non immune stressors than to the pathophysiology of depression, these results are intriguing and worthy of further
investigation.

Cachexia and anorexia are prominent features of chronic infection
and cancer. PBMC isolated from cancer patients release more TNF-α
in response to stimulation than cells from healthy controls. Elevated
serum levels of TNF-α have also been reported in patients with a variety

of malignancies. The possible involvement of TNF-α in the metabolic changes occurring in malnutritional status has been deduced from the demonstration of an enhanced release of this cytokine by stimulated PBMC from short-term acutely starved subjects and non stimulated PBMC from patients afflicted with anorexia nervosa. Anorexia, weight loss and wasting are important causes of morbidity in AIDS patients. The pathophysiology of the AIDS wasting syndrome includes three components: impaired nutrient intake, decreased nutrient absorption and metabolic alterations.[65] The possibility that cytokines play a key role in the AIDS wasting syndrome has been proposed on the basis of two different sets of findings: (i) HIV infection is accompanied by activation of several cytokine genes; and (ii) many characteristic features of the AIDS wasting syndrome are similar to those which are seen in other chronic infections and cancer, and which are due to cytokines. Infection and activation of CD4+ lymphocytes and macrophages by HIV are associated with the production of a range of cytokines, especially TNF-α. Elevated levels of this cytokine are found in the plasma of HIV infected patients. In addition, PBMC of HIV-infected individuals produce higher levels of TNF-α. Other cytokines, including IL-1, IL-6 and TGF-β are also hyperexpressed in HIV-1 infection.

HIV-1 is able to enter the CNS and infect macrophages and microglial cells. However, the small number of infected cells contrasts with the extensive amount of functional alterations, of which the most well known is AIDS-related dementia. Many AIDS patients present a primary neuropsychiatric disorder, including slowness in thinking, forgetfulness, cognitive changes, drowsiness, weakness and difficulties in concentrating. These neurological abnormalities occur in as many as 80% of all AIDS patients, cannot generally be attributed to opportunistic infections and may even precede most other clinical signs of infection with HIV. Since HIV-1 is well known to induce the synthesis of TNF-α, IL-1, IL-6 and IL-2 from microglia, it has been proposed that these cytokines contribute to the neurological symptoms observed in AIDS patients. Examination of brains from HIV-1 infected patients by immunocytochemistry revealed higher expression of IL-1 in endothelial cells and TNF-α in macrophages and microglial cells. TGF-β has also been identified in HIV-1 infected brain and appears to be localized exclusively to tissue areas with pathological abnormalities.

CONCLUSION

Sufficient evidence is now available to accept the concept that cytokines are interpreted by the brain as molecular signals of sickness. Sickness can actually be considered as a motivation, that is, a central state that organizes perception and action in face of this particular threat which is represented by infectious pathogens. A sick individual does not have the same priorities as a well one, and this reorganization of priorities is mediated by the effects of cytokines on a number of peripheral and central targets. The elucidation of the mechanisms that are involved in these effects should give new insight on the way sickness and recovery processes are organized in the brain.

ACKNOWLEDGMENTS

Supported by INSERM, INRA, Université de Bordeaux II, Pôle Médicament Aquitaine, Ministère de l'Environnement, DRET and NIH (MH51569-02 and DK49311).

REFERENCES

1. Hart BL. Biological basis of the behavior of sick animals. Neurosci Biobehav Rev 1988; 12:123-37.
2. Kluger MJ. Fever: its biology, evolution and function. Princeton: Princeton University Press, 1979.
3. Kent S, Bluthé RM, Kelley KW et al. Sickness behavior as a new target for drug development. Trends Pharmacol Sci 1992b; 13:24-28.
4. Tazi A, Dantzer R, Crestani F, Le Moal M. Interleukin-1 induces conditioned taste aversion in rats: A possible explanation for its pituitary-adrenal stimulating activity. Brain Res 1988; 473:369-371.
5. Tazi A, Crestani F, Dantzer R. Aversive effects of centrally injected interleukin-1 are independent of its pyrogenic activity. Neurosci Res Comm 1990; 7:159-165.
6. Segall MA, Crnic LS. A test of conditioned taste aversion with mouse interferon-α. Brain Behav Immun 1990; 4:223-231.
7. Bluthé RM, Dantzer R, Kelley KW. Effects of interleukin-1 receptor antagonist on the behavioral effects of lipopolysaccharide in rat. Brain Res 1992b; 573:318-20.
8. Bluthé RM, Sparber S, Dantzer R. Modulation of the behavioral effects of interleukin-1 in mice by nitric oxide. NeuroReport 1992c; 3:207-209.
9. Bluthé RM, Pawlowski M, Suarez S et al. Synergy between tumor necrosis factor α and interleukin-1 in the induction of sickness behavior in mice. Psychoneuroendocrinology 1994a; 19:197-207.
10. Kent S, Bret-Dibat JL, Kelley KW et al. Mechanisms of sickness-induced decreases in food-motivated behavior. Neurosci Biobehav Rev 1995; 20:171-75.
11. Plata-Salaman CR. Cytokines and ingestive behavior: methods and overview. In: De Souza E, ed, Methods in Neuroscience. Orlando: Academic Press 1993; 17:151-168.
12. Babbini M, Gaiardi M, Bartoletti M. Changes in operant behavior as an index of withdrawal state from morphine in rats. Psychonom Sci 1972; 29:142-144.
13. Bluthé RM, Dantzer R, Kelley KW. Corticotropin releasing hormone is not involved in the behavioral effects of peripherally injected interleukin-1 in the rat. Neurosci Res Commun 1989; 5:149-154.
14. Krueger JM, Pappenheimer JR, Karnovsky ML. The composition of sleep promoting factor isolated from human urine. J Biol Chem 1982; 257:1664-1669.
15. Krueger JM, Obal F Jr, Opp M, Toth L, Johannsen L, Cady AB.

Somnogenic cytokines and models concerning their effects on sleep. Yale J Biol Med 1990; 63:157-172.

16. Opp M, Obal F Jr, Krueger JM. Interleukin-1 alters rat sleep: temporal and dose-related effects. Am J Physiol, Regul Integr Comp Physiol 1990; 260:R52-R58.

17. Aubert A, Vega C, Dantzer R et al. Pyrogens specifically disrupt the acquisition of a task involving cognitive processing in the rat. Brain Behav Immun 1995b; 9:129-48.

18. Schweizer A, Feige U, Fontana A et al. Interleukin-1 enhances pain reflexes. Mediation through increased prostaglandin E2 levels. Agents and Actions 1988; 25:246-251.

19. Ferreira SH, Lorenzetti BB, Bristow AF et al. Interleukin-1β is a potent hyperalgesic agent antagonized by a tripeptide analogue, Nature 1988; 334:698-700.

20. Wiertelak EP, Smith KP, Furness L et al. Acute and conditioned hyperalgic response to illness. Pain 1994; 56;227-234.

21. Fukuoka H, Kawatani M, Hisamitsu T et al. Cutaneous hyperalgesia induced by peripheral injection of interleukin-1β in the rat. Brain Res 1994; 657:133-140.

22. Bolles RC. Species-specific defense reactions and avoidance learning. Psychol Rev 1970; 77:32-48.

23. Miller NE. Some psychophysiological studies of motivation and of the behavioral effects of illness. Bull Brit Psychol Soc 1964; 17:1-20.

24. Aubert A, Goodall G, Dantzer R. Compared effects of cold ambient temperature and cytokines on macronutrient intake in rats. Physiol Behav 1995a; 57:869-73.

25. Katsuura G, Arimura A, Koves K et al. Involvement of organum vasculosum of lamina terminalis and preoptic area in interleukin-1β induced ACTH release. Am J Physiol 1990; 258:E163-E171.

26. Stitt JT. Evidence for the involvement of organum vasculosum laminae terminalis in the febrile response of rabbits and rats. J Physiol 1985; 368:501-511.

27. Layé S, Parnet P, Goujon E et al. Peripheral administration of lipopolysaccharide induces the expression of cytokine transcripts in the brain and pituitary of mice. Mol Brain Res 1994;27:157-62.

28. Parnet P, Amindari S, Wu C et al. Expression of type I and type II interleukin-1 receptors in mouse brain. Mol Brain Res 1994; 27:63-70.

29. Kent S, Bluthé RM, Dantzer R et al. Different receptor mechanisms mediate the pyrogenic and behavioral effects of interleukin-1. Proc Natl Acad Sci USA 1992a; 89:9117-20.

30. Gatti S, Bartfai T. Induction of tumor necrosis factor-α mRNA in the brain after peripheral endotoxin treatment: comparison with interleukin-1 family and interleukin-6. Brain Res 1993; 624:291-295.

31. Van Dam AM, Brouns M, Louisse S, Berkenbosch F. Appearance of interleukin-1 in macrophages and ramified microglia in the brain of endotoxin-treated rats: a pathway for the induction of non specific symptoms of sickness. Brain Res 1992; 588:291-296.

32. Bret-Dibat JL, Kent S, Couraud JY et al. A behaviorally active dose of lipopolysaccharide increases sensory neuropeptides levels in mouse spinal cord. Neuroci Lett 1994; 173:205-209.

33. Wan W, Wetmore L, Sorensen CM et al. Neural and biochemical mediators of endotoxin and stress-induced c-fos expression in the rat brain. Brain Res Bull 1994; 34:7-14.

34. Watkins LR, Wiertelak EP, Goehler LE et al. Neurocircuitry of stress-induced hyperalgesia. Brain Res 1994; 639:283-299.

35. Bluthé RM, Walter V, Parnet P et al. Lipopolysaccharide induces sickness behavior in rats by a vagal mediated mechanism. CR Acad Sci Paris, Sciences de la Vie 1994b; 317:499-503.

36. Bret-Dibat JL, Bluthé RM, Kent S et al. Lipopolysaccharide and interleukin-1 depress food-motivated behavior in mice by a vagal-mediated mechanism. Brain Behav Immun 1995; 9:242-46.

37. Fleshner M, Goehler LE, Hermann J et al. Interleukin-1β-induced corticosterone elevation and hypothalamic NE depletion is vagally mediated. Brain Res Bull 1995; 37:605-610.

38. Watkins LR, Goehler LE, Relton JK et al. Blockade of interleukin-1 induced hyperthermia by subdiaphragmatic vagotomy: evidence for vagal mediation of immune-brain communication. Neurosci Lett 1995; 183:27-31.

39. Layé S, Bluthé RM, Kent S et al. Subdiaphragmatic vagotomy blocks induction of IL-1β mRNA in mice brain in response to peripheral LPS. Am J Physiol, Regul Integr Comp Physiol. 1995; 268:R1327-R1331.

40. Bluthé RM, Crestani F, Kelley KW et al. Mechanisms of the behavioral effects of interleukin-1: Role of prostaglandins and CRF. Ann NY Acad Sci 1992a; 650:268-75.

41. Bluthé RM, Michaud B, Kelley KW et al. Vagotomy attenuates behavioral effects of interleukin-1 injected peripherally but not centrally. Neuroreport 1996; 7:1485-1488.

42. Dantzer R. How do cytokines say hello to the brain? Neural versus humoral mediation, Eur Cytokine Netw 1994; 5:271-73.

43. Uehara A, Sekiya C, Takasugi Y et al. Anorexia induced by interleukin-1: involvement of corticotropin-releasing factor. Am J Phyisol, Regul Integr Comp Physiol, 1989; 257:R613-R617.

44. Del Cerro S, Borrell J. Interleukin-1 affects the behavioral despair response in rats by an indirect mechanism which requires endogenous CRF. Brain Res 1990; 528:162-164.

45. Dunn A, Antoon M, Chapman Y. Reduction of exploratory behavior by intraperitoneal injection of interleukin-1 involves brain corticotropin-releasing factor. Brain Res Bull 1991; 26:539-542.

46. Crestani F, Seguy F, Dantzer R. Behavioral effects of peripherally injected interleukin-1: role of prostaglandins. Brain Res 1991; 542:330-334.

47. Shimizu H, Uehara Y, Shimomura Y et al. Central administration of ibuprofen failed to block the anorexia induced by interleukin-1. Eur J Pharmacol 1991; 195:281-284.

48. Otterness IG, Golden HW, Seymour PA et al. Role of prostaglandins in the behavioral changes induced by murine interleukin 1α in the rat. Cytokine 1991; 3:333-338.

49. Crnic LS, Segall MA. Prostaglandins do not mediate interferon-α effects on mouse behavior. Physiol Behav 1992; 51:349-352.

50. Daun JM, McCarthy DO. The role of cholecystokinin in interleukin-1-induced anorexia. Physiol Behav 1993; 54:237-241.

51. Kluger MJ. Fever: Role of pyrogens and cryogens. Physiol Rev 1991; 71:93-127.

52. Dantzer R, Bluthé RM, Kelley KW. Androgen-dependent vasopressinergic neurotransmission attenuates interleukin-1-induced sickness behavior. Brain Res 1991; 557:115-120.

53. Goujon E, Parnet P, Aubert A et al. Corticosterone regulates behavioral effects of lipopolysaccharide and interleukin-1β in mice. Am J Physiol, Regul Integr Comp Phsyiol 1995a; 269:R154-R159.

54. Goujon E, Parnet P, Cremona S et al. Endogenous glucocorticoids regulate central effects of interleukin-1β on body temperature and behavior in mice. Brain Res 1995b; 702:173-80.

55. Goujon E, Parnet P, Layé S et al. Adrenalectomy enhances pro-inflammatory cytokines gene expression in the spleen, pituitary and brain of mice in response to lipopolysaccharid., Mol Brain Res 1996; 36:53-62.

56. Smith A, Tyrrell D, Coyle K et al. Effects of interferon alpha in man: a preliminary report. Psychopharmacology 1988; 96:414-416.

57. Oitzl MS, van Oers H, Schöbitz B et al. Interleukin-1β, but not interleukin-6, impairs spatial navigation learning. Brain Res 1993; 613:160-163.

58. Gibertini M, Newton C, Friedman H et al. Spatial learning impairment in mice infected with Legionella pneumophila or administered exogenous interleukin-1β. Brain Behav Immun 1995; 9:113-128.

59. Kendell RE. Chronic fatigue, viruses and depression. Lancet 1991; 337:160-62.

60. Chao CC, Janoff EN, Hu S et al. Altered cytokine release in peripheral blood mononuclear cell cultures from patients with the chronic fatigue syndrome. Cytokine 1991; 3:292-295.

61. Goldstein JA. Chronic fatigue syndromes: the limbic hypothesis. Binghamton: Harworth Medical Press, 1993.

62. Maes M, Smith R, Scharpe S. The monocyte-T lymphocyte hypothesis of major depression. Psychoneuroendocrinology, 1995; 20:111-16.

63. Maier SF, Watkins L. Intracerebroventricular interleukin-1 receptor antagonist blocks the enhancement of fear conditioning and interference with escape produced by inescapable shock. Brain Res 1995; 695:279-282.

64. Yirmiya R. Endotoxin produces a depressive-like episode in rats. Brain Res 1996; 711:163-174.

65. Grunfeld C, Kotler DP. Pathophysiology of the AIDS wasting syndrome. AIDS Clin Rev 1992:191-224.

CHAPTER 8

THE ROLE OF CYTOKINES IN NEURODEGENERATION

Nancy J. Rothwell

INTRODUCTION

Cytokines are established mediators and modulators of inflammation and damage in peripheral tissues. However, the realization that they are also involved in degeneration and damage in the nervous system has emerged only recently, but is now an intensely active and exciting field of research.

The term cytokine encompasses diverse groups of molecules, many of which have now been discovered in the nervous system and have been implicated directly or indirectly in neurodegeneration. To cover the whole of this field within a single chapter would necessitate an encyclopedic, and hence rather uninteresting, approach. This chapter will therefore focus on the groups of cytokines known as interleukins (IL), tumour necrosis factors (TNF) and interferons (IFN), but will not discuss the extensive literature on growth factors or neutrophins, except where these relate directly to actions of the above cytokines. Numerous, excellent reviews exist on growth factors, and their involvement in neurodegeneration could form a large and separate book. Indeed, the focus of this subject is further enhanced by the fact that studies to date have centered on a relatively small number of cytokines, mainly the IL-1 family, TNFα, and to a lesser extent IL-6, IL-8 and IFNβ and γ. There is little doubt that within the next few years, information on most of the other cytokines will emerge.

The term neurodegeneration is broad and open to misinterpretation. In this chapter it is used to describe neuronal death or damage which signals the demise of neurones, and thus encompasses both the acute and chronic neurological disorders which feature neuronal loss and/or severe dysfunction due to damage. Here the discussion will be focused largely on degeneration in the brain. The primary examples of

Cytokines in the Nervous System, edited by Nancy J. Rothwell. © 1996 R.G. Landes Company.

these conditions are stroke, head injury, dementias, movement disorders, multiple sclerosis, motor neurone disease and infections which lead to neurodegeneration. Cytokines have now been implicated in all of these. This chapter will discuss the evidence for their involvement, their likely or possible mechanisms of action, and potential avenues for development of therapeutic agents.

CYTOKINE EXPRESSION IS INCREASED IN NEURODEGENERATIVE CONDITIONS

The current literature provides evidence that expression of specific cytokines (IL-1 is the most prominent) is increased in almost every form of experimental or clinical neurodegeneration (see Table 8.1). However, it is important to determine the location, time course, extent and nature of this induction, and most importantly whether cytokine synthesis is primary (and therefore possibly causal), or secondary to damage. The latter question is most relevant since cytokines are frequently induced by tissue damage and inflammation, so it may not be surprising that they are found in diseased brain.

The expression of cytokines in normal brain is usually very low, leading to problems of detection by standard techniques for measuring mRNA or protein. Those cytokines which have been localized in rodent brain (e.g., many neurotrophins, IL-1, IL-6, IL-8, cytokine induced neutrophil chemoattractant (CINC), TNF-α) have usually been found in greatest quantities in the hippocampus, with varying expression in cortex, hypothalamus, cerebellum and sometimes other brain regions (e.g., refs. 1-9). Marked and very rapid increases in synthesis of cytokines have also been reported in brain tissue or cerebrospinal fluid of patients or experimental animals in conditions associated with damage to neurones (e.g., refs. 2-5, 7-18, and see Table 8.1).

Measurements in patients have generally been confined to CSF, in which cytokines are often barely measurable, or post mortem brain tissue. Increased levels of neurotrophins or growth factors, IL-1, IL-6, TNF-α and IFNs have been observed in patients with viral or bacterial infections of the CNS, Alzheimer's disease, Parkinson's disease, multiple sclerosis and epilepsy. However, results of these studies are often variable, and may not reflect accurately local production within brain tissue. More detailed post-mortem analyses of brains from patients with Alzheimer's disease of Down syndrome reveals localized expression of IL-1 associated with amyloid plaques.[12,13,19]

The magnitude and time course of cytokine expression (e.g., IL-1, IL-6, TNF-α, neurotrophins) have been characterized in response to a variety of insults in experimental animals, and found to increase rapidly after systemic or brain injections of excitotoxins, or inflammatory agents such as endotoxin, cerebral ischemia, traumatic brain injury, experimental allergic encephalomyelitis, meningitis, or cerebral malaria (see ref. 20). IL-1β mRNA expression has been detected as early as 15-30 min after induction of forebrain ischemia in the rat[17,21] (and

Table 8.1. Increased cytokine expression in the brain

Experimentally Induced Neurological Disorders

Excitotoxins
Bacterial, viral, parasite infections
Scrapie
Mechanical injury
Stroke
Experimental Allergic Encephalomyelitis

Clinical Neurological Disorders

Brain infections–HIV, meningitis, edema
Brain injury
Stroke
Multiple sclerosis
Alzheimer's disease
Down's Syndrome
Parkinson's disease
Amyotrophic lateral sclerosis

Licinio, Loddick and Rothwell, unpublished data), and cytokine expression can be sustained for many days.[14,22]

The cell source of cytokines in the brain in response to these insults has not been fully established. Glia, especially microglia and astrocytes, produce many cytokines (e.g., IL-1, IL-6, TNF-α, GM-CSF, TGFβ_1 and TGFβ_2) in vitro, and appear to be an important source of cytokines after brain damage (see refs. 23-25). However, other resident brain cells including neurones, endothelial and ependymal cells, as well as invading immune cells may also synthesize cytokines. Macrophages, lymphocytes and neutrophils can invade the CNS after damage, inflammation or breakdown of the blood brain barrier,[26-29] and almost certainly contribute to increased cytokine production in chronic neurological disorders. Significant macrophage and lymphocyte invasion is usually delayed by several hours after damage to the brain, so it is unlikely that they are involved directly in the very early production of cytokines seen after experimental damage. In contrast, neutrophils have been implicated directly in ischemic brain damage, and particularly in reperfusion injury (see below).

EFFECTS OF CYTOKINES RELEVANT TO NEURODEGENERATION

Numerous and diverse actions of cytokines in the nervous system have been identified (see ref. 30). These include effects on specific cell types including neurones, glia and brain endothelial cells, usually identified in vitro, and actions on complex physiological and pathological systems in the brain. These rapidly emerging advances in cytokine neurobiology have, as might be predicted, led to some confusion and questions. Many reports exist of apparently conflicting actions of cytokines; for example, while some observed effects might *promote* neurodegeneration, others are likely to inhibit neuronal death. The considerable task now remains of distinguishing which of the observed effects of exogenous cytokines reflects accurately the roles of the endogenous molecules in vivo. A further complication has been quite frequent discrepancies between findings arising from in vivo and in vitro systems.

In vivo studies on cytokines, usually in rodents, have implicated several of these in neurodegeneration, either as mediators or inhibitors of brain inflammation, damage and neuronal death (see below). Acute injection or infusion of cytokines into the brains of experimental animals does not appear to cause overt neuronal death, but can elicit a number of other responses reminiscent of neurodegenerative disease. Thus, several proinflammatory cytokines such as IL-1, TNF-α, IFNγ and IL-6 can elicit glial activation, cause edema, blood brain barrier damage and expression of potentially detrimental molecules such as MHC class II antigen, eicosanoids, complement, nitric oxide and superoxides (see ref. 30). Chronic overexpression of cytokines such as IL-6 in the brain of transgenic mice (IL-6 gene coupled to a promoter causing expression in astrocytes) leads to severe neurological symptoms, including neurodegeneration and premature death of the animals.[31] These dramatic effects of cytokine overexpression may of course reflect actions during brain development rather than direct effects on mature neurones.

Although acute administration of cytokines into the brain of adult rodents does not cause overt neuronal damage, several pieces of evidence indicate that cytokines can exacerbate neurodegenerative conditions. Clinical trials of IFNγ in multiple sclerosis showed a marked increase in clinical symptoms (see ref. 32). Intracerebral injection of IL-1 markedly enhances neurodegeneration caused by experimentally induced excitotoxic damage[33] (Lawrence and Rothwell, unpublished data) or focal[34] or global[9] cerebral ischemia in rodents. Such exacerbation may reflect non-specific, proinflammatory or pyrogenic actions of IL-1; for example, increased body temperature markedly increases ischemic damage, particularly in global ischemia (see ref. 35). However, although the doses of IL-1 used do cause significant fever, icv injection of IL-6, which has similar pyrogenic potency, inhibits rather than enhances damage caused by focal ischemia (Loddick, LeFeuvre and Rothwell, unpublished data).

Hallenbeck has proposed that increased synthesis of endogenous cytokines is a major risk factor for stroke.[36] Stimulation of endogenous cytokine expression by systemic administration of lipopolysaccharide increases the incidence of stroke in spontaneously hypertensive rats,[36] but a causal link between these events remains to be proven.

INVOLVEMENT OF CYTOKINES IN ACUTE NEURODEGENERATION

It is now clear that cytokines are overexpressed in most forms of neurodegeneration, but more importantly, we now know that in experimental animals, their synthesis *precedes* or *coincides with* progression of neuronal damage and death. For IL-1, IL-8, TNF-α and some growth factors, the time course of expression has been mapped in some detail, and is consistent with their role in neurodegeneration, though such data do not provide direct evidence.

Similarly, studies on the effects of recombinant cytokines in experimental systems in vivo or in vitro, or on animals overexpressing the genes for specific cytokines support, but do not prove, an involvement in neuronal death. Icv injection of small amounts (ng) of recombinant IL-1 or IL-8 markedly exacerbates infarct volume and edema induced by focal cerebral ischemia in the rat.[9,34,37] In the case of IL-8, this action may be due to stimulation of neutrophil invasion into the brain.[9] Both IL-1 and IL-8 are potent pyrogens, which, at the doses used, cause increases in body temperature,[30] which may directly enhance ischemic damage (see above). However, injection of IL-6, which causes a similar increase in body temperature, *inhibits* infarct volume caused by middle cerebral artery occlusion (MCAo) in the rat (see above). In addition to the specific concern about febrile responses, IL-1 and IL-8 are potent proinflammatory molecules, which may influence neuronal damage through effects not directly relevant to the normal course of events after ischemia.

The most convincing evidence supporting the involvement of cytokines in neurodegeneration derives from experimental studies in which the synthesis or action of the endogenous cytokine is inhibited. Icv administration of antibodies which bind to IL-1 or IL-8, markedly reduces edema, neuronal damage and neutrophil invasion caused by reversible ischemia in the rat.[9,37] However, by far the most extensive studies on the involvement of IL-1 in neurodegeneration have used a recombinant preparation of the naturally occurring IL-1 receptor antagonist (IL-1ra), which appears to be a highly selective competitive antagonist of IL-1 receptors.

The first published report on effects of IL-1ra on acute neurodegeneration[38] indicated dramatic (up to 70%) inhibition of infarct volume by icv injection of IL-1ra in rats subjected to focal cerebral ischemia (middle cerebral artery occlusion, MCAo) or excitotoxic (NMDA receptor activation) brain damage. Subsequently, these observations have been verified and extended. IL-1ra, unlike many anti-ischemic agents, protects striatal as well as cortical tissue in permanent MCAo, is effective when administered 30-60 min after MCAo, and protection seems to be sustained for at least seven days.[34] IL-1ra is effective when the gene is delivered into the brain using an adenovirus vector[39] or when injected systemically (iv or ip).[40,41] In the latter case it inhibits IL-1ra and edema, neuronal survival, neutrophil invasion and behavioral impairment in rats in which the MCAo is induced by coagulation or by insertion of a filament via the carotid artery.[40,41]

In addition to its effects on cerebral ischemia, IL-1ra also reduces significantly neuronal damage caused by traumatic brain injury (lateral fluid percussion injury[42]), AMPA or NMDA receptor activation[33,43] in the rat, or heat stroke in the rabbit,[44] and reduces the clinical symptoms of EAE in the rat[45] (and see Table 8.2).

IL-1ra appears to act as an endogenous neuroprotective agent, which is induced rapidly in the brain after traumatic or ischemic brain damage, and inhibition of IL-1ra (by passive immunoneutralization) significantly increases infarct volume resulting from both insults[46] (Toulmond, Loddick and Rothwell unpublished data). Thus the balance between local expression of IL-1 and IL-1ra may be critical in determining neuronal survival or death.

MECHANISMS OF ACTION OF CYTOKINES IN NEURODEGENERATION

At the present time, the exact mechanism(s) of action of cytokines in neurodegeneration is not known, and as results become available,

---

Table 8.2. Neuroprotective effects of inhibiting IL-1

Cerebral Ischemia

Permanent MCAo (rat, mouse)–IL-1ra icv, sub cut, βICE inhibitor, icv
    Reversible MCAo (rat)–IL-antibody, icv
    Perinatal hypoxia (rat)–IL-1ra, sub cut

Brain Injury

    Lateral fluid percussion injury (rat)–IL-1ra icv

Excitotoxic Damage

    Brain infusion–NMDA, AMPA, QA (rat)- IL-1ra, icv
    Systemic–kainic acid (rat)- IL-1ra, sub cut

Heat Stroke

    Rabbit–IL-1ra, sub cut

Brain Inflammation

    EAE (rat)–IL-1ra, IL-1R antibody, systemic

an increasingly complex picture appears to be emerging. Rather than discuss the potential or proposed mechanisms of action of each cytokine (which may in fact share common effects), this review will focus on IL-1, for which most information is available.

ACTIONS ON NEURONES

Few studies have investigated cytokine actions on neurones in vivo. However, our recent data indicate that IL-1/IL-1ra may have indirect actions at distant sites in the brain through neuronal activation. Infusion of IL-1 in the rat striatum markedly exacerbates excitotoxic damage in the cortex, but has no effect on local degeneration in the striatum. This indicates a specific effect of IL-1 in the striatum which leads, indirectly, to distant effects in the cortex.[33,43] Further support for this unexpected finding has been obtained from experiments on cerebral ischemia. In rats subjected to MCAo, injection of IL-1ra into the cerebral ventricles or into the striatum, inhibits striatal or *cortical* degeneration, while cortical infusion of IL-1ra has no effect on local damage or that in the striatum (Stroemer and Rothwell, unpublished data). The nature of these indirect effects of IL-1 are unknown, but may involve synthesis of corticotrophin releasing factor (CRF). IL-1 is a potent inducer of CRF synthesis which mediates many actions of IL-1 in the brain (see ref. 30). CRF mRNA is induced markedly after MCAo or brain trauma, in the amygdala, and to a lesser extent in the cortex.[47] Also, icv injection of a CRF receptor antagonist significantly inhibits neuronal damage caused by MCAo, excitotoxins[48,49] or traumatic brain injury (Roe, McGowan and Rothwell, unpublished data). Thus, IL-1 may activate a neuronal pathway which involves CRF expression and leads to cell death at distinct sites. However, further experiments are required to verify this hypothesis, and IL-1 and CRF may mediate neurodegeneration through independent actions.

In vitro studies on isolated neurones or primary neuronal cultures have failed to identify actions of IL-1 or IL-1ra consistent with their in vivo effects on neurodegeneration. Indeed, IL-1 *protects* cultured neurones against excitotoxins, probably through induction of NGF,[50,51] and other cytokines can also protect neurones.[52] IL-1 also enhances GABA activity, can inhibit calcium currents and has inhibitory actions on LTP, all of which might exert neuroprotective effects (see refs. 30, 53-55). Furthermore, we have failed to observe any effects of IL-1 or IL-1ra on glutamate release in synaptosomes or brain slices (Allan and Rothwell unpublished data).

Together, these observations indicate that IL-1 may be neuroprotective rather than neurotoxic. However, they fail to mimic observations in vivo, indicating that in vitro systems are inappropriate. Many factors differ significantly between the complex systems of the adult brain in vivo and neurones cultured from immature animals in isolation. A primary factor may be the absence of other brain cells, notably glia and endothelial cells.

ACTIONS ON GLIA

Glia (particularly microglia) represent the primary source of many cytokines in the brain and are probably a major target of action (see refs. 56-58). Glia participate in almost all brain functions, and are involved directly in processes of neurodegeneration, e.g., as primary mediators of glutamate uptake, buffers of ionic concentrations and sources of neuroprotective and neurotoxic molecules. IL-1 acts on microglia to produce a variety of neurotoxins (see refs. 57, 59, 60). Furthermore, although IL-1 is *protective* to neurones cultured in the absence of glia, in co-cultures of neurones and glia, IL-1 (and TNF-α) is neurotoxic at low concentrations on effects which may be mediated via release of free radicals (Relton and Newberger, personal communication).

ACTIONS ON ENDOTHELIAL CELLS

Brain endothelial cells are increasingly recognized as contributors to various forms of neurological disease, particularly stroke. IL-1 and TNF-α have a variety of direct and indirect actions on the brain endothelium, for example causing nitric oxide release and blood brain barrier damage (see refs. 53-55, 61).

Peripheral immune cells (lymphocytes, neutrophils and macrophages) which normally have limited access to the CNS, enter the brain parenchyma after injury, infection or ischemia, however their direct contribution to neuronal degeneration is uncertain. Recent data implicate neutrophil invasion in experimental neurodegeneration resulting from reversible cerebral ischemia.[28,62-64] IL-8 can stimulate tissue invasion of neutrophils directly,[65] while IL-1 may stimulate invasion indirectly through induction of adhesion molecules, such as ICAM-1, on the brain endothelium.

INVOLVEMENT OF CYTOKINES IN CHRONIC NEURODEGENERATION

Increased cytokine expression (usually measured in CSF or postmortem brain tissue) is a frequent observation in chronic neurological disorders such as multiple sclerosis, Alzheimer's disease and Parkinson's disease (Table 8.2). The most obvious explanation for these observations, as for acute conditions, is that cytokines are induced as a *result* of the tissue damage and/or inflammation associated with these conditions, since most cytokines are produced by damaged cells. *Proof* of a *direct* involvement of cytokines in chronic neurological disorders is limited by the availability of valid experimental models.

Several animal models of multiple sclerosis have been used widely. In particular, experimental autoimmune encephalomyelitis (EAE) in the rat, induced by immunization against myelin basic protein resulting in autoimmune attack on oligodendrocytes, shows clinical symptoms similar to multiple sclerosis, but over a shorter time scale. Inhibition of TNF-α or IL-1 actions markedly inhibits the clinical symptoms of EAE in the rat,[66] and the success of IFN β in human multiple sclerosis has been ascribed to its inhibition of synthesis and/or actions of IFNγ, IL-1 and TNF-α (see ref. 32). Thus, it is likely that cytokines are important in the development and progression of multiple sclerosis, though their site(s) and mechanisms of action have not been fully elucidated and may reside in the periphery.

EVIDENCE FOR A ROLE OF CYTOKINES IN ALZHEIMER'S DISEASE

Several pieces of evidence support the hypothesis that cytokines (particularly IL-1) participate in the development and/or progression of Alhzeimer's disease:

- Overexpression of IL-1 has been reported in Alzheimer's disease and Down syndrome.
- IL-1, and possibly other cytokines, can elicit most of the responses which characterize Alzheimer's disease.
- IL-1 is involved in acute neurodegeneration which may share common mechanisms with those involved in Alzheimer's disease.
- IL-1 is involved in damage caused by head injury, which is a significant risk factor associated with Alzheimer's disease.
- Increasing evidence is emerging for inflammatory responses in the brains of patients with Alzheimer's disease, and preliminary studies suggesting benefits of anti-inflammatory drugs.
- Some data suggest general alterations in immune function in patients with Alzheimer's, including changes in circulating cytokines or their production by peripheral immune cells.

A number of studies have reported increased expression of cytokines, particularly IL-1, in the brain of patients with Alzheimer's disease compared to age-matched controls. These measurements, usually based on immunocytochemistry or ELISA, reveal overexpression of IL-1α and IL-1β (for example, see refs. 10-13) predominantly in senile plaques, with the greatest IL-1α immunoreactivity in diffuse amyloid plaques showing a profusion of β-APP positive dystrophic plaques.[13]

IL-1 appears to be largely microglial in origin, and is found particularly in brain regions associated with Alzheimer-type pathology, i.e., frontal, parietal and tempral cortex, hippocampus and thalamus, with little of no increase in expression in cerebellum.[11] Other cytokines, most notably IL-6, are also reportedly elevated in Alzheimer brains.[67,68] In contrast to these studies using immunocytochemistry, Wood et al[69] reported no difference in mature IL-1β or IL-1ra in the brains of Alzheimer patients detected by ELISA, but observed significantly raised IL-6 concentrations, though this difference may reflect methodological discrepancies.

Hubermenan et al[70] observed that mononuclear cells from patients with Alzheimer's disease showed greater IL-2 and IFNγ secretion than those from controls, but lower production of IL-1β. In contrast, Singh[70a] described increased production of IL-1, IL-2 and IL-6 by cells from demented patients, while Pirtilla et al[71] found that IL-1β was not elevated in cerebrospinal fluid or serum of Alzheimer patients. Fillit et al[72] observed increased circulating TNF-α in Alzheimer's disease. These apparent anomalies are not surprising in view of the fact that many cytokines (and particularly IL-1) are released into circulation infrequently, transiently and usually in low quantities. These molecules are generally produced and act locally within tissues, though reported changes in cytokine production by immune cells may reflect a fundamental change in sensitivity to stimuli which elicit cytokine production.

The reported changes in brain cytokine expression, though generally consistent, also need to be treated with some caution. Neuronal damage, glial activation and inflammation are common features of Alzheimer's disease, (see refs. 67, 68, 73) which may *cause* cytokine expression in the brain. Such expression may therefore be secondary to the disease, but could nevertheless exacerbate progression. A second, potentially important factor, rarely considered, is the fact that demented patients are at higher risk of infections (many of which may not be readily detected). Systemic infection induces brain cytokine expression (see ref. 20), and could therefore contribute to the findings in Alzheimer patients.

Notwithstanding these caveats about cytokine overexpression and a causal role for these molecules in disease initiation and/or progression, the data described above are highly consistent with the proposal that cytokines are involved, at least in some way in Alzheimer's disease. Further evidence for this hypothesis derives from the fact that cytokines (of which IL-1 is the most studied) can induce responses in vivo or in vitro which characterize Alzheimer pathology. Of these responses, perhaps the most notable is the potent effect of IL-1 on induction of β-APP, and perhaps amyloid itself,[74,75] from a number of cells or cell lines in culture, including neurones and glia.[74,76,77,78] This effect is consistent with the fact that β-APP is considered an acute phase protein which is highly responsive to cytokines such as IL-1 and IL-6. The APP gene contains cytokine repsonse elements and β-APP mRNA expression is also induced in cultured mouse neuronal or glial cells by IL-2, IL-3, IL-6, NGF, PGF or GM-CSF, but surprisingly not by TNF-α[77], though the bioactivity and species specificity of TNF-α may reflect this negative result.

PHARMACOLOGICAL APPROACHES TO CYTOKINE MODULATION

Experimental studies to date have generally used large molecules (e.g., rIL-1ra or neutralizing antibodies) to modulate cytokine action.

These molecules, together with antibodies to receptors, soluble receptors or naturally occurring binding proteins are currently under development for systemic modulation of cytokine actions in a number of disease states. However, as large molecules, some of which may be immunogenic, they are not ideally suited for the treatment of CNS disorders such as stroke. A primary goal therefore is to develop non-peptide inhibitors of cytokine action through a number of approaches.

Some cytokines, of which IL-1β is most notable, require enzymatic cleavage from an inactive precursor. Thus, inhibitors of ICE activity may be beneficial in the treatment of stroke. Several cysteine protease inhibitors have now been developed, which can inhibit IL-1 release, and in vivo responses associated with IL-1 action (e.g., refs. 80, 81). However, in view of the role of related enzymes in apoptosis, it may be necessary to develop strategies which specifically inhibit IL-1β release, because even if apoptosis is an important event in ischemic brain damage, chronic inhibition of this process could lead to undesirable side effects such as tumor development, and ICE inhibitors may be particularly unsuitable for chronic treatments.

Attempts to develop non-peptide antagonists of cytokine receptors have not yet proven fruitful, but might offer a selective approach and modulation of signal transduction pathways may also be of therapeutic benefit. However, for most cytokines, the signal transduction pathways in the brain are not known, although recent studies suggest the involvement of novel MAP kinases and the transcription factors AP-1 and NF+β in IL-1 signalling (see ref. 82). Several naturally occurring inhibitors of IL-1 synthesis or action have been identified (see Table 8.3) including other cytokines, lipocortin, αMSH and arginine vasopressin[83-86] (and Loddick, Toulmond and Rothwell, unpublished data). As lipocortin and IL-1ra are known to be endogenous neuroprotective agents[46,86] (and Loddick, Toulmond and Rothwell, unpublished data), it is possible that other endogenous IL-1 inhibitors may be of similar benefit in ischemic brain damage.

SUMMARY AND CONCLUSIONS

The evidence that cytokines mediate or inhibit acute neurodegeneration in experimental animals is now considerable. This direct modification of their synthesis or actions may offer an attractive route for developing therapeutic targets.

There is also now considerable information implicating cytokines in acute and chronic clinical conditions associated with neurodegeneration. This evidence rests largely on experimental studies on animals, with no information from direct intervention in humans, with the notable exception of multiple sclerosis. Extrapolation from studies in animals to humans depends on the relevance of the experimental models which have yet to be proven since no treatment has yet been proven successful in patients with acute chronic neurodegeneration,

---

Table 8.3. Endogenous inhibitors of inflammatory cytokines

Anti-inflammatory Cytokines

IL-1ra, IL-4, IL-10, IL-13, TGFβ

Cytokine Inhibitors

Soluble receptors, binding proteins
αMSH, αMSH fragments
Glucocorticoids
Lipocortin-1
Vasopressin

---

and for some conditions such as Alzheimer's disease, relevant animal models are not yet available. Nevertheless, it is likely from the available information, which is accumulating exponentially, that cytokines are likely to be critical factors in many forms of neurodegenerative disease and provide exciting candidates for therapeutic intervention.

REFERENCES

1. Breder C, Dinarello CA, Saper CB. Interleukin-1 immunoreactive innervation of the human hypothalamus. Science 1988; 240:321-324.
2. Liu T, McDonnell PC, Young PR, White RF, Siren AL, Hallenbeck JM, Barone FC, Feuerstein GZ. Interleukin-1 beta mRNA expression in ischemic rat cortex. Stroke 1993; 24:1746-1750.
3. Liu T, Clark RK, McDonnell PC, Young PR, White RF, Varone FC, Feuerstein GZ. Tumor necrosis factor-α expression in ischemic neurons. Stroke 1994; 25:1481-1488.
4. Liu T, Young PR, McDonnell PC, White RF, Barone FC, Feuerstein GZ. Cytokine-induced neutrophil chemoattractant mRNA expressed in cerebral ischemia. Neurosci Lett 1993; 164:125-128.
5. Romero LI, Schettini G, Lechan RM, Dinarello CA, Reichlin S. Bacterial lipopolysaccharide induction of IL-6 in rat telencephalic cells is mediated in part by IL-1. Neuroendocrinol. 1993; 57:892-897.
6. Schobitz B, de Kloet ER, Sutanto W, Holsboer F. Cellular localisation of interleukin 6 mRNA and interleukin 6 receptor mRNA in rat brain. Eur J Neurosci 1993; 5:1426-1435.
7. Wang XK, Yue TL, Young PR, Barone FC, Feuerstein GZ. Expression of interleukin-6, c-fos, and zif268 mRNAs in rat ischemic cortex. J Cereb Blood Flow Metab 1995; 15:166-171.
8. Wießner C, Gehrmann J, Lindholm D, Topper R, Kreutzberg GW, Hossman KA. Expression of transforming growth factor-β1 and interleukin-1β mRNA in rat brain following transient forebrain ischemia. Acta Neurophathol 1993; 86:439-446.

9. Yamasaki Y, Matsuo Y, Matsuura N, Onodera H, Itoyama Y, Kogure K. Transient increase of cytokine-induced neutrophil chemoattractant, a member of the interleukin-8 family, in ischemic brain areas after focal ischemia in the rat. Stroke 1995; 26:318-323.

10. Cacabelos R, Alvarez XA, Fernandez-Novoa, L Franco A, Mangues R, Pellicer A, Nishimura T. Brain interleukin-1β in Alzheimer's disease and vascular dementia. Methods Find Exp Clin Pharmacol 1994; 16:141-51.

11. Cacabelos R, Barquero M, Garcia P, Alvarez XA, Varela E. Cerebrospinal fluid interleukin 1β (IL-1β) in Alzheimer's disease and neurological disorders. Methods Find Exp Clin Pharmacol 1991; 13:455-458.

12. Griffin WS, Stanley LC, Ling C, White L, MacLeod V, Perrot LJ, White CL, Araoz C. Brain interleukin 1 and S-100 immunoreactivity are elevated in Down syndrome and Alzheimer's disease. Proc Natl Acad Sci USA 1989; 86:7611-7615.

13. Griffin WS, Sheng JG, Roberts GW, Mrak RE. Interleukin- expression in different plaque types in Alzheimer's disease:significance in plaque evolution. J Neuropathol Exp Neurol 1995; 54:276-281.

14. Ianotti F, Kida S, Weller R, Buhagier G, Hillhouse E. Interleukin-1β in focal cerebral ischemia in rats. J Cereb Blood Flow and Metab 1993; 13:s125.

15. McClain CJ, Cohen D, Ott L, Dinarello CA, Young B. Ventricular fluid interleukin-1 activity in patients with head injury. J Lab Clin Med 1987; 110:48-54.

16. Merrill JE, Chen IS. HIV-1, macrophages, glial cells and cytokines in AIDS nervous system disease. FASEB J 1991; 5:2391-2397.

17. Minami M, Kuraishi Y, Yabuuchi K, Yamazaki A, Satoh M. Induction of interleukin-1 mRNA in rat brain after transient forebrain ischemia. J Neurochem 1992; 58:390-392.

18. Taupin V, Toulmond S, Serrano A, Benavides J, Zavala F. Increase in IL-6, IL-1 and TNF levels in rat brain following traumatic lesion. J Neuroimmunol 1993; 42:177-186.

19. Goldgaber D, Harris HW, Hla T, Caciagi T, Donnelly RJ, Jacobson JS, Vitek MP, Gajdusek DC. Interleukin-1 regulates synthesis of amyloid precursor protein nRNA in human endothelial cells. Proc Natl Acad Sci 1989; 86:7606-7610.

20. Hopkins SJ, Rothwell NJ. Cytokines and the nervous system 1:expression and recognition. TINS 1995; 18:83-88.

21. Yabuuchi K, Minami M, Katsuata S, Yamasaki A, Satoh A. An in situ hybridisation study of interleukin-1 beta induced by transient forebrain ischemia in the rat brain. Mol Brain Res 1994; 26:135-142.

22. Buttini M, Sauter A, Boddeke HWGM. Induction of interleukin-1 beta messenger RNA after focal cerebral ischemia in the rat. Mol Brain Res 1994; 23:126-134.

23. Banati RB, Gehrmann J, Schubert P, Kreutzberg GW. Cytotoxicity of microglia. Glia 1993:7:111-127.

24. Fontana A, Kriestensin F, Dubs R, Gemsa D, Weber E. Production of

prostaglandin E and interleukin-1 like factor by cultured astrocytes and C6 glioma cells. J Immunol 1982; 129:2413-2419.

25. Woodroofe MN, Sarna GS, Wawda M, Hayes GM, Loughlin AJ, Tinker A, Cuzner ML . Detection of interleukin-1 and interleukin-6 in adult rat brain following mechanical injury by in vivo microdialysis:evidence of a role for microglia in cytokine production. J Immunol 1993; 33:227-236.

26. Chopp M, Zhang RL, Chen H, Li Y, Jiang N, Rusche JR. Post ischemic administrations of an anti Mac-1 antibody reduces ischemia cell damage after transient middle cerebral artery occlusion in rats. Stroke 1994; 25:869-876.

27. Kochanek PM, Hallenbeck JM. Polymorphonuclear leukocytes and mono-cytes/macrophages in the pathogenesis of cerebral ischemia and stroke. Stroke 1992; 23:1367-1379.

28. Lindsberg PJ, Siren AL, Feuerstein GZ, Hallenbeck JM. Postischemic antagonism of neutrophil adherence has an acute thereapeutic effect on functional recovery in the deteriorating stroke model in rabbits. J Cereb Blood Flow Metab 1991; 11:s754.

29. Perry VH, Andersson P-B, Gordon S. Macrophages and inflammation in the central nervous system. TINS 1993; 16:268-273.

30. Rothwell NJ, Hopkins SJ. Interactions between cytokines and the ner-vous system II:Actions and mechanisms. TINS 1995; 18:130-136.

31. Campbell IL, Abraham CR, Masliah E, Kemper P, Inglis JD, Oldstone MBA, Mucke L. Neurological disease induced in transgenic mice by cere-bral overexpression of interleukin-6. Proc Natl Acad Sci USA 1993; 90:10061-10065.

32. Panitch AS. Interferons in multiple sclerosis. Drugs 1992; 44:946-962.

33. Allan SM, Lawrence CB, Rothwell NJ. Mechanism of action of interleukin-1 (IL- 1) in excitotoxic brain damage in the rat. Soc Neurosci Abst 1995; 21:37.14.

34. Loddick SA, Rothwell NJ. Neuroprotective effects of human recombinant interleukin-1 receptor antagonist in focal cerebral ischemia in the rat. J Cereb Blood Flow Metab 1996; In press.

35. Ginsberg MD, Globus MYT, Dietrich WD, Busto R. Temperature modu-lation of ischemic brain injury—a synthesis of recent advances. In:Kogure K, Hossman KA, Siesjo BK, eds. Progress in Brain Research. Elsevier 1993; 13-22.

36. Hallenbeck JM, Dutka AJ, Vogel SN, Heldman E, Doron DA, Feuerstein. Lipopolysaccharide-induced production of tumor necrosis factor activity in rats with and without risk factors for stroke. Brain Res 1991; 541:115-120.

37. Yamasaki Y, Matsuura N, Shozuhara H, Onodera H, Itoyama Y, Kogure K. Interleukin-1 as a pathogenetic mediator of ischemic brain damage in rats. Stroke 1995; 26:676-681.

38. Relton JK, Rothwell NJ. Interleukin-1 receptor antagonist inhibits ischemic and excitotoxic neuronal damage in the rat. Brain Res Bull 1992; 29:43-46.

39. Betz A, Yang G-Y, Davidson BL. Attenuation of stroke size in rats using

an adenoviral vector to induce overexpression of interleukin-1 receptor antagonist in brain. J Cereb Blood Flow Metab 1995; 15:547-551.

40. Garcia JH, Liu K-F, Relton JK. Interleukin-1 receptor antagonist decreases the number of necrotic neurones in rats with middle cerebral artery occlusion. Am J Pathol 1995; 147:1477-1486.

41. Relton JK, Martin D, Thompson RC, Russel DA. Peripheral administration of interleukin-1 receptor antagonist inhibits brain damage after focal cerebral ischemia in the rat. Exp Neurol 1996; 138: 206-212.

42. Toulmond S, Rothwell NJ. Interleukin-1 receptor antagonist inhibits neuronal damage caused by fluid percussion injury in the rat. Brain Res 1995; 671:261-266.

43. Lawrence CB, Rothwell NJ. Interleukin-1 receptor antagonist inhibits NMDA and AMPA receptor induced brain damage in the rat. Br J Pharmacol 1994; 112:484P.

44. Lin MT, Kao TY, Jin YT, Chen CF. Interleukin-1 receptor antagonist attenuates the heat stroke-induced neuronal damage by reducing ischemia in rats. Brain Res Bull 1995; 37:595-598.

45. Martin D, Chinookoswong N, Miller G. The interleukin-1 receptor antagonist (rhIL-1ra) protects against cerebral infarction in a rat model of hypoxia-ischemia. Exp Neurol 1995; 130:362-367.

46. Toulmond S, Rothwell NJ. Time-course of IL-1 receptor antagonist (IL-1ra) expression after brain trauma in the rat. Soc Neurosci Abst 1995; 21:200.2.

47. Wong M-L, Loddick SA, Bongiorno PB, Gold PW, Rothwell NJ, Licinio J. Focal cerebral ischemia induces CRH mRNA in rat cerebral cortex and amygdala. Neuro Report 1995; 6:1785-1788.

48. Lyons MK, Anderson RE, Meyer FB. Corticotrophin releasing factor antagonist reduces ischemic hippocampal neuronal injury. J Cereb Blood Flow Metab 1991; 545:339-342.

49. Strijbos PJLM, Relton JK, Rothwell NJ. Corticotrophin-releasing factor antagonist inhibits neuronal damage induced by focal cerebral ischemia or activation of NMDA receptors in the rat brain. Brain Res 1994; 656:405-408.

50. Strijbos PJLM, Rothwell NJ. Interleukin-1 attenuates excitatory amino acid induced neurodegeneration in vitro:involvement of nerve growth factor. J Neurosci 1995; 15:3468-3474.

51. Rothwell NJ, Strijbos PJLM. Cytokines in neurodegeneration and repair. Int J Devl Neurosci 1995; 13:179-185.

52. Araujo DM. Contrasting effects of specific lymphokines on the survival of hippocampal neurones in culture:In: Meyer EM ed. The Treament of Dementias. Plenum Press, 1992; 113-122.

53. Rothwell NJ, Lawrence CB, Loddick SA, Strijbos PJLM, Toulmond S. Cytokines and cerebral ischemia. In: Krieglstein J, Oberpichler-Schwenk H. Pharmacology of Cerebral Ischemia 1994; 419-425. Wissenschaftliche Verlagsgesellschaft.

54. Rothwell N, Loddick S, Lawrence C. Cytokines and neurodegeneration.

In: Rothwell NJ. Immune Responses in the Nervous System. 1995: 77-79. BIOS Scientific.

55. Rothwell NJ, Relton JK. Involvement of interleukin-1 and lipocortin-1 in ischemic brain damage. Brain Metab Rev 1993; 5:178-198.

56. Beneviste EN. Cytokines:influence on glial cell gene expression and function. In: Blalock JE, ed. Neuroimmunoendocrinology. Karger: 1992; 106-153.

57. Giulian D. Reactive glia as rivals in regulating neuronal survival. Glia 1993; 7:102-110.

58. Merrill JE. Effects of interleukin-1 and tumor necrosis factor- on astrocytes, microglia, oligodendrocytes, and glial precursors in vitro. Dev Neurosci 1991; 13:130-137.

59. Piani D, Frei K, Do KQ, Cuenod M, Fontana A. Murine brain macrphages induce NMDA receptor mediated neurotoxicity in vitro by secreting glutamate. Neurosci Lett 1991; 133:159-162.

60. Piani D, Spranger M, Frei K, Schaffner A, Fontana A. Macrophage-induced cytotoxicity of N-methyl-D-aspartate receptor positive neurons involves excitatory amino acids rather than oxygen intermediates and cytokines. Eur J Immunol 1992; 22:2429-2436.

61. Quagliarello VJ, Wispelwey B, Long WJ, Scheld WM. Recombinant human interleukin-1 induces meningitis and blood-brain barrier injury in the rat. J Clin Invest 1991; 87:1360-1366.

62. Chen H, Chopp M, Bodzin G. Neutropenia reduces the volume of cerebral infarction after transient middle cerebral artery occlusion in the rat. Neurosci Res Commun 1992; 11:93-99.

63. Shiga Y, Onodera H, Kogure K, Yamasaki Y, Yashima Y, Szozuhara H, Sendo F. Neutrophil as a mediator of ischemic oedema formation in the brain. Neurosci Lett 1991; 125:110-112.

64. Takeshima R, Kirsch JR, Koehler RC, Gomoll AW, Traystman RJ. Effect of neutrophil monoclonal antibody on infarct volume following transient focal cerebral ischemia in cats. J Cereb Blood Flow Metab 1991; 11:s752.

65. Andersson PB, Perry VH, Gordon S. Intracerebral injection of proinflammatory cytokines or leucocyte chemotaxins induces minimal myelomonocytic cell recruitment to the parenchyma of the CNS. J Exp Med 1992; 176:255-259.

66. Martin D, Near SL. Protective effect of the interleukin-1 receptor antagonist (IL-1ra) on experimental allergic encephalomyelitis in rats. J Neuroimmunol 1995b; 61:241-245.

67. Bauer J, Strauss S, Schreiter-Gasser, U Ganter, U, Schlegel P, Will I, Yolk B, Berger M. Interleukin-6 and β-2-macroglobulin indicate an acute-phase state in Alzheimer's disease cortices. FEBS Lett 1991; 285:111-114.

68. Bauer J, Ganter U, Strauss S, Stadmuller G, Frommberger U, Bauer H, Bolk B, Berger M. The participation of interleukin-6 in the pathogenesis of Alzheimer's disease. Res Immunol 1992; 143:650-657.

69. Wood JA, Wood, PL, Ryan, R, Graff-Radford NR, Pilapil, C, Robitaille Y, Quirion R. Cytokine indices in Alzheimer's temporal cortex:no changes

in mature IL-1 beta or IL-1ra but increases in the associated acute phase proteins IL-1, alpha 2-macroglobulin and C-reactive protein. Brain Res 1993; 629:245-252.

70. Huberman M, Shalit F, Roth-Deri I, Gutman B, Brodie C, Kott E, Sredni B. Correlation of cytokine secretion by mononuclear cells of Alzheimer patients and their disease stage. J Neuroimmunol 1994; 52:147-152.

70a. Singh VK. Studies of neuroimmune markers in Alzheimer's disease. Mol Neurobiol 1994; 9:73-81.

71. Pirtilla T, Hehta PD, Frey H, Wisniewski HM. Alpha 1-antichymotrypsin and IL-1 beta are not increased in CSF or serum in Alzheimer's disease. Neurobiol Aging 1994; 15:313-317.

72. Fillit H, Ding W, Buee L, Kalman J, Altsteil L, Lawlor B, Wolf-Klein G. Elevated circulating tumour necrosis factor levels in Alzhiemer's disease. Neurosci Lett 1991; 129:318-320.

73. Vandenabeele P, Fiers W. Is amyloidogenesis during Alzheimer's disease due to an IL-1/IL-6 mediated acute phase response in the brain? Immunol Today 1991; 12:207-219.

74. Buxbaum JD, Ruefli AA, Parker CA, Cypress, AM, Greengard P. Calcium regulates processing of the Alzheimer amyloid protein precursor in a protein kinase C-independent manner. Proc Natl Acad Sci USA 1994; 91:4489-4493.

75. Forloni G, Demicheli F, Giorgi S, Bendotti C, Angeretti N. Expression of amyloid precursor protein mRNAs in endothelial, neuronal and glial cells; modulation by interleukin-1. Mol Brain Res 1992; 16:128-134.

76. Dash PK, Moore AN. Enhanced processing of APP induced by IL-1 beta can be reduced by indomethacin and nordihydroguaiaretic acid. Biochem Biophys Res Commun 1995; 208:542-548.

77. Ohyagi Y, Tabira T. Effect of growth factors and cytokines on expression of amyloid beta protein precursor mRNAs in cultured neural cells. Mol Brain Res 1993; 18:127-132.

78. Vasilakos JP, Carroll RT, Emmerling MR, Doyle PD, Davis RE, Kin KS, Shivers BD. Interleukin-1 beta dissociates beta-amyloid precursor protein and beta-amyloid peptide secretion. FEBS Lett 1994; 354:289-292.

79. Chapman KT. Synthesis of a potent reversible inhibitor of interleukin-1β converting enzyme. Inorganic Medic Chem Letts 1992; 2:613-618.

80. Elford PR, Heng, R, Revesz L, MacKenzie AR. Reduction of inflammation and pyrexia in the rat by oral adminstration of SDZ 224-015, an inhibitor of the interleukin-1β converting enzyme. Br J Pharmacol 1995; 115:601-606.

81. Fletcher DA, Agarwal L, Chapman KT, Chin J, Egger LA, Limjuco G, Luell S, MacIntyre DE, Peterson EP, Thornberry NA, Kotsura MJ. A synthetic inhibitor of interleukin-1β production in vitro and in vivo. J Int Cyt Res 1995; 15:243-248.

82. O'Neill LAJ. Towards an understanding of the signal transduction pathways for interleukin-1. Biochim Biophys Acta 1995; 1266:31-44.

83. Dubois CM, Ruscetti FW, Palaszynki EW, Falk LA, Oppenheim JJ, Keller

JR. Transforming growth factor β is a potent inhibitor of interleukin-1 (IL-1) receptor expression:proposed mechanism of inhibition of IL-1 action. J Exp Med 1990; 172:737-744.

84. Henrich-Noack P, Prehn JHM, Krieglstein J. Neuroprotective effects of TGF-β1. J Neural Transm 1994; 43:33-45.

85. Quirion R, Araujo DM, Lapchak PA, Seto D, Chabot JG. Growth factors and lymphokines:modulators of cholinergic neuronal activity. Can J Neurol Sci 1991; 18:390-393.

86. Relton JK, Strijbos PJLM, O'Shaughnessy CT, Carey F, Forder RA, Tilders FJH, Rothwell NJ. Lipocortin-1 is an endogenous inhibitor of ischemic damage in the rat brain. J Exp Med 1991; 174:305-310.

CHAPTER 9

CYTOKINES AS THERAPEUTIC AGENTS IN NEUROLOGICAL DISORDERS

D. Martin, J.K. Relton, G. Miller, A. Bendele, N. Fischer
and D. Russell

THERAPEUTIC APPLICATIONS FOR DRUGS MODULATING CYTOKINE SYSTEMS

Drugs which modulate cytokine systems may have therapeutic applications in diseases of the central nervous system (CNS) such as stroke, multiple sclerosis (MS), head trauma and Alzheimer's disease (AD). In this review the emphasis will be on interleukin-1 (IL-1), tumor necrosis factor (TNF) and interferons (INF), since there is substantial preclinical and clinical data to support a role for these cytokines in these neurological disorders.

MULTIPLE SCLEROSIS

Multiple sclerosis (MS) is a chronic disease of the CNS that is unpredictable and often crippling. It is unpredictable because its course can be relatively benign, somewhat disabling, or devastating. It is crippling because communication between the CNS and other parts of the body is disrupted, at worst leaving the person unable to speak, walk or write. It is estimated that between 200,000 and 500,000 people have MS in the United States. Most people with MS are diagnosed between the ages of 30 and 50. Women are almost twice as likely to develop MS as men; whites more than twice as likely as other races.[1,2]

Multiple sclerosis is primarily a demyelinating disease in which an inflammatory cell infiltrate is a characteristic pathologic feature of active

Cytokines in the Nervous System, edited by Nancy J. Rothwell. © 1996 R.G. Landes Company.

lesions within the CNS and loss of neurologic function. The cause of MS is unknown, but two major hypotheses have been proposed: (1) the initiating event is an infection which leads to cellular infiltration in response to the pathogen; alternatively, (2) the initial infiltrating cells are autoimmune in nature and recognize CNS specific proteins that are displayed by local antigen presenting cells. Neither hypothesis has been proven, but regardless of the initiating event, inflammatory cells play a major role in the resulting pathology.[1,2]

Several lines of evidence have implicated cytokines in the inflammatory demyelinating process.[2-4] Cytokines such as interleukin-1 (IL-1), tumor necrosis factor (TNF) and interferon-gamma (IFNγ) are released by inflammatory cells (macrophages and lymphocytes) in the periphery or within the CNS. These cells have the potential to enhance inflammation in brain and spinal tissue by promoting myelin destruction by macrophages in inflamed CNS tissue, resulting in demyelination. Furthermore, resident brain cells, microglia and astrocytes, also produce cytokines which may augment the inflammatory process.[5-8] Much of the evidence supporting a role for cytokines in MS has come from studies using selective antagonists and agonists. These results are discussed in greater detail below.

INTERFERONS

The rationale for investigating interferons as potential disease-modifying agents in MS arose from: (1) their antiviral and immunomodulatory properties; (2) their limited toxicity during clinical use; and (3) the observation that cells isolated from patients with MS have deficient production of type I (interferon alpha and beta, IFN-α and IFN-β) and type II (gamma, IFN-γ) interferons.[9] In clinical trials recombinant IFN-1b (an 18.5 kDa synthetic analogue of recombinant human interferon beta) reduced significantly both the rate of MS-related exacerbations and the number of new lesions detected on serial magnetic resonance imaging. These findings led to the approval of this agent in 1993 for the treatment of ambulatory relapsing-remitting MS patients in the US.[10]

Although the mechanisms by which IFNβ-1b produces therapeutic benefits are unknown, several drug-related activities may be relevant to the immunopathogenesis of MS. For example, this interferon inhibits viral replication and is a potent antiproliferative agent. Among its immunomodulatory activities are the ability to increase the cytotoxicity of natural killer cells; phagocytic activity of macrophages; antibody dependent cytotoxicity of polymorphonuclear leucocytes and killer cells; expression and shedding of human leucocyte antigens and tumor-associated antigens and suppressor cell function in vitro.[11-15] Furthermore, the drug reduces IFN-γ secretion by activated lymphocytes, production of tumor necrosis factor by macrophages, and expression of IFN-γ-induced class II major histocompatibility antigens on antigen-presenting glial cells.[9,16,17]

IFN-β-1b has been approved only for ambulatory relapsing-remitting patients between the ages of 18-50. However, there is no reason to assume that there is any biological difference between these individuals and MS patients with relapsing-progressive clinical course or those who are older than 50 years.[10] The main concern is that the occurence of negative side effects may not justify treatment with this agent in the less severely affected MS population. In general treatment with IFN-β-1b is well tolerated, however, side effects include: injection site pain, fatigue, allergic reaction, nausea, headache, cardiac arrhythmia, abnormal liver enzyme levels and flu-like symptoms.[10]

Apart from exploiting the possibilities offered by regulatory cytokines, one could try to inhibit proinflammatory cytokines with antibodies or soluble receptors. Several such compounds exist, but of particular interest are the TNF and IL-1 inhibitors. Although inhibitors of TNF or IL-1 are not presently in clinical trials for MS, they are for the autoimmune disease, rheumatoid arthritis (RA).

INTERLEUKIN-1

Results from preclinical and clinical studies have implicated IL-1 and TNF as possible mediators of chronic neuroinflammatory diseases. For example, IL-1 can augment the in vitro activation of encephalitogenic T lymphocytes and enhance adoptive transfer of EAE.[18] Other studies demonstrated increased IL-1 immunoreactivity in astrocytes from EAE animals.[19,20] In addition, increased levels of IL-1β in cerebrospinal fluid have been detected in guinea pigs suffering from chronic relapsing EAE.[21] The critical animal studies used two types of IL-1 inhibitors: the recombinant human interleukin-1 receptor antagonist (rhIL-1ra) and a soluble mouse recombinant IL-1 receptor (sIL-1R).[22-24] In active rat EAE studies the IL-1 antagonists reduced the clinical severity of the disease.[23,25] In addition, clinical studies have shown increased levels of IL-1 in CSF from MS patients.[26,27] Furthermore, increased IL-1 mRNA expression has been detected in the perivascular inflammatory cuffs from CNS tissue of MS patients.[28] Taken together, these data suggest that IL-1 may initiate or promote inflammation within the CNS.

TUMOR NECROSIS FACTOR

Tumor necrosis factors (TNFs) form a unique family of cytokines that are very pleiotropic in nature.[29] The family is composed of TNF-α, lymphotoxin alpha (LT-α), and LT-β. Recently, both in vitro and in vivo studies have suggested a role for TNFs in the pathology of MS. Interest in TNF as a possible mediator of MS pathology arose from in vitro studies. Application of TNF-α was demonstrated to be cytotoxic to murine oligodendrocytes and caused demyelination of cultured neurons.[30,31] Further studies suggested an apoptotic mechanism of oligodendrocyte injury evoked by TNF.[32]

A typical feature of MS pathology is reactive gliosis, where reactive astrocytes constitute the major cell type in demyelinating plaques. In vitro studies have suggested that TNFs are mitogenic for astrocytes.[33] Thus TNFs may contribute to the development of astrocyte proliferation in MS plaques.

TNF and lymphotoxin were detected by immunocytochemistry in acute and chronic active MS lesions but not in silent lesions.[34] It was shown that immunoreactivity for both cytokines was noted on cells at the edge of the lesion, but cellular localization was different. LT-α was localized to CD3 positive lymphocytes and microglial cells, whereas TNF-α was associated with astrocytes and activated foamy macrophages.[26,27,35,36] Furthermore, disease progression was correlated with high levels of TNF and soluble TNF receptor (sTNF-R) in the cerebrospinal fluid. The finding that high levels of TNF-α in the CSF of chronic-progressive MS patients did not correspond with serum levels of TNF-α suggested that this factor is produced locally within the CNS.[35,36]

A reasonable hypothesis based on these results is that inhibition of TNF may alleviate the tissue damage, clinical signs, and/or demyelination associated with this disease. The effects of TNF inhibition have been tested in experimental autoimmune encephalomyelitis (EAE), an acute or chronic relapsing inflammatory demyelinating disease of the CNS resulting from passive immunization with MBP-sensitized T cell clones or active immunization with MBP. Neutralizing antibodies specific to both TNF-α and TNF-γ or only to TNF-α prevented the adoptive transfer of EAE with sensitized T cell clones in SJL/J mice.[37,38] Treatment with the human recombinant type 1 soluble receptor of TNF (rsTNF-R1), which binds TNF and acts as an inhibitor, also prevented the adoptive transfer of EAE in passively immunized mice.[39,40] Furthermore, a dimeric polyethylene glycol linked form of the type 1 soluble receptor of TNF, PEG-(rsTNF-R1)$_2$ (now referred to as TNFbp) significantly reduced the clinical and pathological sequelae of actively acquired EAE in rats.[41] Taken together these studies suggest that TNF plays a pivotal role in the mediation of inflammatory events that lead to demyelination and neurologic dysfunction. Therefore, direct inhibition of TNF in MS patients with a recombinant soluble receptor of TNF or other inhibitors may limit the inflammatory sequelae associated with this neurodegenerative disorder.

It is interesting to speculate that combination therapies with cytokines or other agents such as growth factors may be potentially very effective in ameliorating clinical symptoms as well as perhaps enhancing central nervous system remyelination. Furthermore, combination therapies which lower individual drug doses, may also reduce some of the side effects currently associated with IFN-β–1b treatment.[10]

CEREBRAL ISCHEMIA AND RELATED DISORDERS

Cerebral ischemia treatment has recently become a major area for clinical trials, because of the medical need (250,000-500,000 patients per year), and the potential profits for pharmaceutical companies. Ischemia related conditions are not only fatal, but leave thousands of patients mentally and physically disabled.[42,43]

Each year thousands of patients are enrolled in therapeutic trials, testing the hypothesis that thrombolytics or other classes of agents such as glutamate antagonists, free radical scavengers, or calcium channel blockers will improve functional outcome. Besides the thrombolytic agents (which is under review by the U.S. Food and Drug Administration for approval), none of the above compounds have been approved for use in stroke or head trauma. Although the basic research in many instances is well founded, the main reasons for failure to develop new classes of therapeutics, besides the obvious efficacy and/or toxicity issues, has been trial design. Whereas most trials in the 1970s and 1980s enrolled patients as late as 24 or even 48 hours after stroke onset, at the present time it would be almost unthinkable to extend the time of inclusion beyond the "magic limit" of 6 hours.[44] There are several reasons for this: First, experiments conducted in the nonhuman primate that led to the concept of the ischemic penumbra clearly documented that "time is brain", with apparently only little penumbra remaining beyond 3 hours.[45,46] Second, thus far many clinical trials have reported negative, only marginally significant, and/or positive but poorly reproducible findings; late inclusion of patients was considered to be responsible for treatment failure.[47,48] Third, recent studies have demonstrated that it is possible to have stroke patients seen in the hospital within 3 to 6 hours after onset.[49,50] Fourth, the success of intravenous theraputic thrombolysis in less than 6 hours after myocardial infarction stimulated a similar design for acute stroke.[51]

If these variables can be controlled it may be possible to evaluate novel therapeutic agents for ischemia related conditions, with the primary endpoint being improved functional outcome. The following sections will discuss the rationale for using cytokine inhibitors and, in particular, the interleukin-1 receptor antagonist (IL-1ra) in stroke and/or head trauma conditions based on recent preclinical animal studies.

EVIDENCE FOR CYTOKINE PRODUCTION IN THE ISCHEMIC BRAIN

Focal ischemia presents clinically as stroke and is commonly modelled by reversible or irreversible occlusion of one middle cerebral artery (MCAo). It is now evident that IL-1β, TNF and IL-6 are involved in pathogenic mechanisms resulting from various forms of ischemia in the CNS.[52] These cytokines are detected in brain tissue of rats within 6-8 h after MCAo and IL-1β mRNA was induced within 15 min of forebrain ischemia in the rat.[53,54] The very rapid expression precedes invasion of peripheral immune cells, and might originate from injured neurons, microglia or perivascular cells. Furthermore, the increases in IL-1β mRNA has been shown to persist for several days.[54]

TNF-α is present in the brain and can be synthesized and released by astrocytes, microglia and some neurons.[5-7,55] Increased expression of this cytokine has been demonstrated after intraperitoneal injection of the neurotoxin kainic acid, mechanical injury, fetal hypoxia and MCAo.[53,54,56-58] Although TNF expression appears to be increased in these studies, there isn't any direct evidence demonstrating that TNF-α causes neuronal cell loss in cerebral ischemia. However, recent studies have shown that peripheral administration of TNFbp, reduced infarct volume and attenuated the inflammatory processes resulting from middle cerebral artery occlusion in rats (John Hallenbeck, personal communication).

IL-6 is produced by cells of the central nervous system in response to other cytokines, viral challenge or brain damage.[7,8,56,59-61] Experimental models of mechanical injury or subarachnoid hemorrhage in the rat results in significant upregulation of IL-6 mRNA and the IL-6 receptor in brain tissue.[56,60] In stroke patients, CSF levels of IL-6 are markedly elevated and have been shown to correlate with both infarct volume and measures of stroke outcome.[62,63] Unlike IL-1β and TNF-α, IL-6 levels are elevated in the serum of stroke patients. The close correlation between circulating IL-6 levels and the extent of brain infarction has led to the proposal that this cytokine could be used as a clinical indicator of the severity of brain damage.[62,63]

EFFECTS OF CYTOKINE INHIBITORS ON BRAIN ISCHEMIA

Inhibitors of IL-1, in particular IL-1ra, has been assessed in rodent models of focal ischemia. Most of these studies involved surgical occlusion of the middle cerebral artery, a clinically relevant experimental procedure for induction of cerebral ischemia, followed by histological assessment of the resulting cerebral infarct after survival times ranging from 24 h to 7 days. In control animals this procedure leads to massive necrosis of cortical and striatal tissue, with up to half of the total cerebral cortex in the affected hemisphere showing signs of ischemic damage. IL-1ra given centrally or peripherally at the time of occlusion or shortly after provided a remarkable degree of neuroprotection, with the volume of cortical and striatal tissue damage in rat and mouse reduced by approximately one-half.[64-68] Furthermore, improvements in neurologic outcome and a reduction in inflammatory cell infiltrate into the damaged areas were also observed in IL-1ra treated animals.[67] The therapeutic window in these models ends between 1 and 2 hours post occlusion, therefore delaying drug administration in these models dramatically reduces efficacy. However, other ischemia related models have shown longer therapeutic windows, and IL-1ra has been examined in some of these models. For example, seven day old rats subjected to unilateral carotid artery occlusion and subsequent hypoxia exhibited severe brain damage, which could be attenuated by IL-1ra given up to 3 hrs after the start of hypoxia.[69] These data indicate that endogenous IL-1 is a mediator of cerebral ischemia and that modification of its actions may be of therapeutic benefit.

HEAD TRAUMA

It is a startling revelation to note that mortality from traumatic brain injury over the past 12 years has exceeded the cumulative number of American battle deaths in all wars since the founding of this country. The total number of head injuries is conservatively estimated at over 2 million per year, with 500,000 serious enough to require hospital admission. The outlook for many survivors is poor. Approximately 70,000 to 90,000 each year face a life-long debilitating loss of function, and increased susceptibility for epilepsy and Alzheimer's disease. Furthermore, the overall economic cost to society approaches $25 billion each year with little hope of improvement.[70]

Head trauma can be modelled in the rat by fluid percussion delivered to the cerebral cortex via trephine hole. Outcome is assessed by neurological testing and histopathology. Head trauma models have shown increases in IL-1β mRNA and IL-1 protein in rat brain, suggesting that IL-1β may play a role in the post-traumatic pathologic sequelae of brain injury.[61,71] Additional studies, using IL-1ra in a rat head trauma model reduced the infarct volume by approximately 50%, confirming the role of IL-1 in ischemia related conditions.[72] Toxins that are produced by microglia and/or macrophages may act in part directly or indirectly via IL-1 receptor activation and result in injury at time points that are delayed considerably beyond the initial trauma.[73] It is possible that the therapeutic time window in head injury is very much greater than cerebral ischemia based on animal studies demonstrating that IL-1ra retains efficacy in head trauma models when administration is delayed by several hours.[72] This factor, combined with the greater rapidity of transport of patients to emergency care facilities and the greater tolerance of cognitive or behavioral side effects (in a comatose patient in intensive care), makes traumatic injury probably the best target for the cerebroprotective effect of IL-1 receptor antagonists.

ALZHEIMER'S DISEASE

Alzheimer's disease (AD) is an age-related, progressive neurologic disorder leading to dementia and is characterized by senile plaques, neurofibrillary tangles and cerebrovascular amyloidosis.[74,75] The filaments of the β-amyloid protein are the major component of the senile plaques and are deposited in cortical and meningeal blood vessels. β-Amyloid peptide is derived from a larger transmembrane glycoprotein precursor known as amyloid precursor protein (APP).[74,75]

AD is one of the most prevalent health problems in western countries and ranks only behind cardiovascular disease, cancer and cerebrovascular disorders in its importance. If preventive measures are not found, the numbers can be expected to increase dramatically as the aging population increases. Although progress has been made on many fronts, a fundamental understanding of the etiology and pathogenesis of AD remains elusive. Recent evidence suggests that neuroimmune mechanisms may play a role in the pathogenesis of AD and attempts to show how these mechanisms may be related to other pathological events in the AD brain, particularly neurodegeneration and β-amyloid protein deposition are being examined.[75]

The role of inflammation in AD has recently gained prominence, and examination of cytokine systems is receiving much attention. Pro-inflammatory cytokines such as IL-1, IL-6 and TNF may be involved in AD.[76,94,95] For example AD and Down's syndrome brains contain 30-fold more interleukin-1-producing microglia and astrocytes as compared to age-matched controls.[76] At the transcriptional level, interleukin-1β increases APP mRNA in primary cultures of astrocytes, cortical neurons and endothelial cells.[77-79] Recent work with primary rat cortical cultures has shown that IL-6 enhances APP mRNA expression in neurons and that β-amyloid fragment 25-35 increases expression of IL-1β in astrocytes.[79,80] Perhaps not surprisingly, the promotor of the APP gene contains heat shock, IL-1 and neuropoietic cytokine-responsive elements. Furthermore, Griffin et al[76] demonstrated that IL-1 immunoreactivity is elevated in tissue sections from patients with AD in brain regions where astrocyte proliferation and astrogliosis are prominent. Moreover, early plaque formation is associated with an increase of activated microglia whereas mature plaques contain fewer microglia. This scenario is confirmed and extended by Huell et al[81] who found that IL-6 is present in the early stages of plaque formation and is correlated with clinical dementia. Further studies have shown an increase in IL-1 levels in the CSF from AD patients and that the increased IL-1 levels correlated with mental deterioration.

Analysis of the various plaque morphologies in AD has shown a progression that implicates IL-1 secreting microglia involvement at the early stages of plaque formation. These results support a model in which microglia are activated at the initial stages of APP deposition and the associated cytokine expression in microglia promotes further APP production and processing, leading to plaque maturation (directly or indirectly stimulating neuritic tangle formation as well). Besides increasing APP synthesis these cytokines are known to stimulate astrocyte proliferation, another feature of AD lesions. Activated astrocytes also produce an array of cytokines which probably contribute to the neuronal degeneration associated with neuritic plaques and dementia.[5,82,83] Taken together these findings suggest that cytokines (at least IL-1) and inflammation become key therapeutic targets for the treatment of AD.

Further support for anti-inflammatory therapy for AD patients stems from several retrospective studies suggesting that prior history of anti-inflammatory drug usage may be inversely associated with frequency of AD symptoms. For example, it was demonstrated that anti-inflammatory drug use is a significant predictor of which of two elderly identical twins will fail to develop AD or develop AD later than the other.[84] Furthermore, rheumatoid arthritis patients (presumably taking anti-inflammatory medication) were approximately 5-fold less likely to carry a secondary diagnosis of AD compared to national statistics for the same age population.[85-87] Retrospective correlations, of course, carry with them many caveats, but it is interesting to speculate that inhibitors of inflammation might have a beneficial effect on attenuating the inflammatory component of AD. Based on the present involvement of IL-1 in AD, modification of IL-1 synthesis or action might prove a valuable area for future therapeutic intervention for AD.

Although this chapter has dealt only with four of the major neurological diseases, there are other conditions in which cytokine receptor systems may contribute to neurologic symptoms and/or pathology including: neuropathic pain, spinal trauma, stress, depression, cerebral malaria and Parkinson's disease,[88-93] (see chapters by Rothwell et al, and Perkins).

Results obtained from the above studies support the hypothesis that cytokine receptor systems play an important role in neurodegenerative disorders. The ability of certain cytokine antagonists when given peripherally or centrally in animal models to attenuate the neurodegenerative processes, suggests potential use of these compounds in the treatment of human neurodegenerative diseases. It remains to be seen whether IL-1 or TNF inhibitors can be safely tolerated in clinical use for neurological disorders like MS, AD or ischemia related conditions. Whatever the outcome in the clinic, the current burgeoning knowledge of cytokine systems will add much to our understanding of the basic brain mechanisms in normal and diseased states within the coming years.

REFERENCES

1. Waksman BH, Reynolds WE. Multiple sclerosis as a disease of immune regulation. Proc Soc Exp Biol Med 1984; 175:282-94.
2. Hafler DA, Weiner HL. T cells in multiple sclerosis and inflammatory central nervous system disease. Immunol Rev 1987; 100:307-32.
3. Burger D, Dayer J-M. Inhibitory cytokines and cytokine inhibitors. Neurology 1995; 45:S39-S43.
4. Brosnan CF, Cannella B, Battistini L et al. Cytokine localization in multiple sclerosis lesions: Correlation with adhesion molecule expression and reactive nitrogen species. Neurology 1995; 45:S16-S21.
5. Chung IY and Benveniste EN. Tumor necrosis factor-α production by astrocytes: Induction by lipopolysaccharide, INF-gamma and IL-1β. J Immunol 1990; 144:2999-3007.

6. Lieberman AP, Pitha PM, Shin HS et al. Production of tumor necrosis factor and other cytokines by astrocytes stimulated with lipopolysaccharide or a neurotropic virus. Proc Natl Acad Sci 1989; 86:6348-52.

7. Lee SC, Liu W, Dickson DW et al. Cytokine production by human fetal microglia and astrocytes: Differential induction by lipopolysaccharide and IL-1 beta. J Immunol 1993; 150:2659-2667.

8. Sebire G, Emilie D, Wallon C et al. In vitro production of IL-6, IL-1β and tumor necrosis factor-α by human embryonic microglial and neural cells. J Immunol. 1993; 150:1517-1523.

9. Weinstock-Gutterman B, Ransohoff RM, Kinkel RP et al. The interferons: biological effects mechanisms of action, and use in multiple sclerosis. Annals of Neurology 1995; 37:7-15.

10. Lublin FD, Whitaker JN, Eidelman BH et al. Management of patients receiving interferon beta-1b for multiple sclerosis: Report of a consensus conference. Neurology 1996; 46:12-18.

11. Taylor JL, Sabran Jl, Grossbey SE. The cellular effects of interferon In: Came PE, Carter WA eds. Interferons and Their Applications. New York: Springer-Verlag, 1984;169-204.

12. Baron S, Tyiring SK, Fleischmann WR Jr et al. The interferons, mechanisms of action and clinical applications. JAMA 1991; 266:1375-1383.

13. Dianzani F, Antonelli G. Physiological mechanism of production and action of interferons in response to viral infections. Adv Exp Med Biol 1989; 257:47-60.

14. Nelson BE and Borden EC. Interferons: biological and clinical effects. Semin Surg Oncol 1989; 5:391-401.

15. Baron S, Dianzani F, Stanton GJ et al. The interferon system: a current review to 1987. Austin, TX: University of Texas, 1987.

16. Rudick RA, Carpenter CS, Cookfair DL et al. In vitro and in vivo inhibition of nitrogen-driven T-cell activation by recombinant interferon beta. Neurology 1993; 43:2080-2087.

17. Lehmann PV, Sercarz EE, Forsthuber T et al. Determinant spreading and the dynamics of the autoimmune T-cell repertoire. Immunol Today 1993; 14:203-208.

18. Mannie MD, Dinarello CA, Patterson, PY. Interleukin-1 and myelin basic protein synergistically augment adoptive transfer activity of lymphocytes mediating experimental autoimmune encephalomyelitis in Lewis rats. J Immunol 1987; 138:4229-4235.

19. Bauer J, Berkenbosch F, Van Dam AM et al. Demonstration of interleukin-1 beta in Lewis rat brain during experimental allergic encephalomyelitis by immunocytochemistry at the light and ultrastructural level. J Neuroimmunol 1993; 48:13-21.

20. Rubio N, Capa L. Differential IL-1 synthesis by astrocytes from Theiler's murine encephalomyelitis virus-susceptible and resistant strains of mice. Cell Immunol 1993; 149:237-247.

21. Symons JA, Bundick RV, Suckling AJ et al. Cerebrospinal fluid interleukin 1 like activity during chronic relapsing experimental allergic encephalo-

myelitis. Clin Exp Immunol 1987; 68:648-654.

22. Eisenberg SP, Evans RJ, Arend NP et al. Primary structure and functional expression from complementary DNA of a known interleukin-1 receptor antagonist. Nature 1990; 343:341-346.

23. Jacobs CA, Baker PE, Roux ER et al. Experimental autoimmune encephalomyelitis is exacerbated by IL-1β and suppressed by soluble IL-1 receptor. J Immunol 1991; 146:2983-2989.

24. Dinarello CA, Thompson RC. Blocking IL-1: interleukin 1 receptor antagonist in vivo and in vitro. Immunol Today 1991; 12:404-410.

25. Martin D, Near SL. Protective effect of the interleukin-1 receptor anatgonist (IL-1ra) on experimental allergic encephalomyelitis in rats. J Neuroimmunol 1995; 61:241-245.

26. Tsukada N, Mijagi K, Matsuda M et al. Tumor necrosis factor and interleukin in the CSF and sera of patients with multiple sclerosis. J Neurol Sci 1991; 102:230-234.

27. Hauser SL, Doolittle TH, Lincoln R et al. Cytokine accumulations in CSF of multiple sclerosis patients. Frequent detection of interleukin-1 and tumor necrosis factor but not interleukin-6. Neurology 1990; 40:1735-39.

28. Woodroofe MN, Cuzner ML. Cytokine mRNA expression in inflammatory multiple sclerosis lesions: detection by non-radioactive in situ hybridization. Cytokine 1993; 5:583-588.

29. Tracey KJ, Wei H, Mannogue KR et al. Cachectin/tumor necrosis factor induces cachexia, anemia and inflammation. J Exp Med 1988; 167:1211-27.

30. Robbins DS, Shirazi Y, Drysdale B-E. Production of cytotoxic factor for oligodendrocytes by stimulated astrocytes. J Immunol 1987; 139:2593-97.

31. Selmaj KW, Raine CS. Tumor necrosis factor mediates myelin and oligodendrocyte damage in vitro. Ann Neurol 1988; 23:339-346

32. Selmaj KW, Raine CS. Experimental autoimmune encephalomyelitis: Immunotherapy with anti-tumor necrosis factor antibodies and soluble tumor necrosis factor receptors. Neurology 1995; 44:S44-S49.

33. Selmaj K, Farooq M, Norton WT et al. Proliferation of astrocytes in vitro in response to cytokines: A primary role for tumor necrosis factor. J Immunol 1990; 144:129-135.

34. Selmaj KW, Raine CS, Cannella B et al. Identification of lymphotoxin and tumor necrosis factor in multiple sclerosis lesions. J Clin Invest 1991; 87:949-954.

35. Beck JP, Rondot L, Catinot E et al. Increased production of interferon gamma and tumor necrosis factor precedes clinical manifestation in multiple sclerosis: Do cytokines trigger off exacerbations? Acta Neurol Scand 1988; 78:318-323.

36. Sharief MK, Hentges R. Association between tumor necrosis factor and disease progression in patients with multiple sclerosis. N Engl J Med 1991; 325:467-472.

37. Selmaj KW, Raine CS, Cross HA. Anti-tumor necrosis factor therapy abrogates autoimmune demyelination. Ann Neurol 1991; 30:694-700.

38. Ruddle NH, Bergmen CM, McGrath ML. An antibody to lymphotoxin and tumor necrosis factor prevents transfer of experimental allergic encephalomyelitis. J Exp Med 1990; 172:1193-1200.

39. Selmaj KW, Papierz W, Glabinski A et al. Prevention of chronic relapsing experimental autoimmune encephalomyelitis by soluble TNF receptor I. Can J Neurol Sci 1993; 20:S157-S161.

40. Hale KK, Smith CG, Baker SL et al. Multifunctional regulation of the biological effects of TNF-α by soluble type I and type II TNF receptors. Cytokine 1995; 7:26-38.

41. Martin D, Near SL, Bendele A et al. Inhibition of tumor necrosis factor is protective against neurologic dysfunction after active immunization of Lewis rats with myelin basic protein. Exp Neurol 1995; 131:221-228.

42. Sacco RL. Risk factors and outcomes for ischemic stroke. Neurology 1995; 45:S10-S14.

43. Dobkin B. The economic impact of stroke. Neurology. 1995; 45:S6-S9.

44. Baron JC, von Kummer R, del Zoppo GJ Treatment of acute ischemic stroke. challenging the concept of a rigid and universal time window. Stroke 1995; 26:2219-2221.

45. Astrup J, Siesjo BK, Symon L. Thresholds in cerebral ischemia: the ischemic penumbra. Stroke 1981; 12:723-725.

46. Jones TH, Morawetz RB, Crowell RM et al. Thresholds of focal cerebral ischemia in awake monkeys. J Neurosurg 1981; 54:773-782.

47. del Zoppo GJ, Zeumer H, Harker LA. Thrombolytic therapy in stroke: possibilities and hazards. Stroke. 1986; 17:595-607.

48. Adams HP, Brott TG, Crowell RM et al. Guidelines for the management of patients with acute ischemic stroke: statement for healthcare professionals from special writing group of the Stroke Council, American Heart Association. Stroke 1994; 25:1901-1914.

49. Fieschi C, Argentino C, Lenzi GL et al. Clinical and instrumental evaluation of patients with ischemic stroke within six hours. J Neurol Sci 1989; 91:311-322.

50. Haley EC, Brott TG, Sheppard GL et al. Pilot randomized trial of tissue plasminogen activator in acute ischemic stroke. Stroke 1993; 24:1000-1004.

51. Zahger D, Gotsman MS. Thrombolysis in the era of randomized trials. Curr Opin Cardiol 1995; 10:372-380.

52. Rothwell NJ, Hopkins SJ Cytokines and the nervous system II: actions and mechanisms of action. TINS 1995; 18:130-136.

53. Minami M, Kurasishi Y, Yamaguchi T et al. Convulsants induce interleukin1β messenger RNA in rat brain. Biochem, Biophys Res Commun 1990; 17:832-837.

54. Liu T, Clark RK, McDonnell PC et al. Tumor necrosis factor-α expression in ischemic neurons. Stroke 1994; 25:1481-1488.

55. Sawada M, Kondo N, Suzumura A et al. Production of tumor necrosis factor-alpha by microglia and astrocytes in culture. Brain Res 1989; 491:394-397.

56. Taupin V, Toulmond S, Serrano A et al. Increase in IL-6 and IL-1 and

TNF levels in rat brain following traumatic brain lesion: influence of pre- and post-traumatic treatment with Ro5 4864, a peripheral-type (p site) benzodiazepine ligand. J Neuroimmunol 1993; 42:177-181.

57. Tchelingerian J-L, Quinonero J, Booss J et al. Localization of TNF-α and IL-1β immunoreactivities in striatal neurons after surgical injury to the hippocampus. Neuron 1993; 10:213-224.

58. Goodman JC, Robertson CS, Grossman RG et al. Elevation of tumor necrosis factor in head injury. J Neuroimmunol 1990; 30:213-217.

59. Frei K, Malipiero UV, Leist TP et al. On the cellular source and function of interleukin-1 produced in the central nervous system in viral diseases. Eur J Immunol 1989; 19:689-694.

60. Matheisen T, Andersson B, Loftenius et al. Increased interleukin-6 levels in cerebrospinal fluid following subarachnoid hemorrhage. J Neurosurg 1993; 78:562-567.

61. Yan HQ, Banos MA, Herregodts P et al. Expression of interleukin IL-1, IL-6 and their respective receptors in the normal rat brain after injury. Eur J Immunol 1992; 22:2963-2971.

62. Beamer NB, Coull BM, Clark WM et al. Interleukin-1 and interleukin-1 receptor antagonist in acute stroke. Ann Neurol 1995; 37:800-804.

63. Tarkowski E, Rosengren L, Blomstrand C et al. Early intrathecal production of interleukin-6 predicts the size of brain lesion in stroke. Stroke 1995; 26:1393-1398.

64. Rothwell LA, Davies C, Fotheringham A et al. Involvement of IL-1 in ischaemic brain damage in the mouse. Soc Neurosci 1995; 25:91.3.

65. Relton JK, Rothwell NJ Interleukin-1 receptor antagonist inhibits ischemic and excitotoxic neuronal damage in the rat. Brain Res Bull 1992; 29:243-246.

66. Relton JK, Martin D, Thompson RC et al. Peripheral administration of interleukin-1 receptor antagonist inhibits brain damage after focal cerebral ischemia in the rat. Exp Neurol 1996; 138:206-213.

67. Garcia JH, Liu KF, Relton JK. Interleukin-1 receptor antagonist decreases the number of necrotic neurons in rats with middle cerebral artery occlusion. Am J Pathol 1995; 147:1477-1485.

68. Loddick SA and Rothwell NJ Neuroprotective effects of recombinant interleukin-1 receptor antagonist in focal cerebral ischaemia in the rat. J Cereb Blood Flow Metabol 1996: In press.

69. Martin D, Chinookoswong N, Miller G. The interleukin-1 receptor antagonist (rhIL-1ra) protects against cerebral infarction in a rat model of hypoxia-ischemia. Exp Neurol 1994; 130:362-367.

70. Gopinath SP, Narayan RK. Clinical treatment: Head Injury. In: Salzman SK, Faden AI, eds. The Neurobiology of Central Nervous System Trauma. Oxford: Oxford University Press, 1994:319-328.

71. Fan L, Young PR, Barone FC et al. Experimental brain injury induces expression of interleukin-1β mRNA in the rat brain. Molecular Brain Res 1995; 30:125-130.

72. Toulmond S, Rothwell NJ. Interleukin-1 receptor antagonist inhibits neu-

ronal damage caused by fluid percussion injury in the rat. Brain Res 1995; 671:261-266.

73. Giulian D. Brain inflammatory cells, neurotoxins, acquired immunodeficiency syndrome. In: Simon RP, ed. Excitatory Amino Acids. New York: Thieme Medical Publishers, Inc., 1992; 229-34.

74. Selkoe DJ. Physiological production of the β-amyloid protein and the mechanism of Alzheimer's disease. Trends Neurosci 1993; 16:403-409.

75. Schenk DB, Rydel RE, May P et al. Therapeutic Approaches related to amyloid-β peptide and Alzheimer's disease. J Med Chem 1995; 38:4141-54.

76. Griffin WST, Stanley LC, Ling C et al. Brain interleukin 1 and S-100 immunoreactivity are detected in Down Syndrome and Alzheimer's disease. Proc Natl Acad Sci USA 1989; 86:7611-7615.

77. Goldgaber D, Harris HW, Hla T et al. Interleukin 1 regulates synthesis of amyloid β-protein mRNA expression within hippocampal neuronal subpopulations in Alzheimer's disease. Proc Natl Acad Sci USA 1989; 86:7606-7610.

78. Forloni G, Demicheli F, Giorgi S et al. Expression of amyloid precursor protein mRNAs in endothelial neuronal and glial cells: modulation by interleukin-1. Mol Brain Res 1992; 16:128-134.

79. Del-Bo R, Angeretti N, Lucca E et al. Reciprocal control of inflammatory cytokines, IL-1 and IL-6, beta-amyloid production in cultures. Neurosci Lett 1995; 188:70-74.

80. Araujo DM, Cotman CW. β-Amyloid stimulates glial cells in vitro to produce growth factors that accumulate in senile plaques in Alzheimer's disease. Brain Res 1992; 569:141-145.

81. Huell M, Strauss S, Volk B et al. Interleukin-6 is present in early stages of plaque formation and is restricted to the brains of Alzheimer's disease patients. Acta Neuropathol Berl 1995; 89:544-551.

82. Giulian D, Baker TJ, Shih N et al. Interleukin-1 of the central nervous system produced by amoeboid microglia. J Exp Med 1986; 164:594-604.

83. Mrak RE, Sheng JG, Griffin WS. Glial cytokines in Alzheimer's disease: review and pathogenic implications. Hum Pathol 1995; 26:816-23.

84. Breitner JC, Gau BA, Welsh KA et al. Inverse association of anti-inflammatory treatments and Alzheimer's disease: initial results of a co-twin control study. Neurology 1994; 44:227-32.

85. McGeer PL, McGeer EG, Rogers J et al. Antiinflammatory drugs and Alzheimer's disease. Lancet 1990; 335:1037-1038.

86. Rogers J, Kirby LC, Hempelman SR et al. Clinical trial of indomethacin in Alzheimer's disease. Neurology 1993; 43:1609-1611.

87. Bannwarth B, Netter P, Lapicque F et al. Plasma and cerebrospinal fluid concentrations of indomethacin in humans. Relationship to analgesic activity. Eur J Clin Pharmacol 1990; 38:343-346.

88. Maier SK, Wiertelak EP, Martin D et al. Interleukin-1 mediates the behavioral hyperalgesia produced by lithium chloride and endotoxin. Brain Res 1993; 623:321-324.

89. Watkins LR, Wiertelak EP, Goehler LE et al. Characterization of cytok-

ine-induced hyperalgesia. Brain Res 1994; 654:15-26.

90. Safieh-Garabedian B, Poole S, Allchorne A et al. Contribution of Interleukin-1β to the inflammation-induced increase in nerve growth factor levels and inflammatory hyperalgesia. Brit J Pharmacol 1995; 115:1265-75.

91. Eling WMC and Kremsner PG. Cytokines in malaria, pathology and protection. Biotherapy. 1994; 7:211-221.

92. Mogi M, Harada M, Kondo T et al. Interleukin-1β, interleukin-6, epidermal growth factor and transforming growth factor-β are elevated in the brain from parkinsonian patients. Neurosci Lett 1994; 180:147-150.

93. Takao T, Hashimoto K, De-Souza EB. Modulation of interleukin-1 receptors in the neuro-endocrine-immune axis. Int J Dev Neurosci 1995; 13:167-78.

94. Bauer J, Ganter U, Strauss S et al. The participation of interleukin-6 in the pathogenesis of Alzheimer's disease. Res Immunol 1992; 143:650-657.

95. Woodroofe MN. Cytokine production in the central nervous system. Neurology 1995; 45:S6-S10.

CYTOKINES AND NOCICEPTION

M.N. Perkins and A.J. Davis

By far the majority of studies with cytokines relate to neuroendocrine functions, fever and sleep.[1] Recently, however, the possibility of a role for cytokines in nociception has been explored. In particular the work of Watkins et al[2,3] have studied central and peripheral involvement of IL-1β in nociception in depth. This chapter will attempt to review the literature relating to those cytokines (primarily IL-1β, IL-2, IL-6, TNF-α and NGF) for which there is some direct evidence for a role in nociception.

CYTOKINES AND NOCICEPTION IN THE CNS

There are relatively few studies addressing the possibility of an involvement of cytokines in nociceptive processing at a central level. Studies which have been conducted have focused on IL-1β and, to a lesser extent, TNF-α. This section will, therefore, concentrate primarily on these cytokines.

BEHAVIORAL STUDIES WITH CENTRALLY ADMINISTERED CYTOKINES

Some of the most direct evidence for an involvement of cytokines in nociception are derived from studies employing administration of cytokines or their antagonists into the brain. The literature is, however, somewhat contradictory with evidence for both pro- and anti-hyperalgesic actions of cytokines.

IL-1β when given icv in rats has been shown to produce thermal hyperalgesia and this is most likely by a central mechanism as similar doses given peripherally failed to induce hyperalgesia.[4] More recently, a similar thermal hyperalgesia was reported following IL-6 injection icv.[5] Interestingly, this IL-6-induced hyperalgesia appeared to differ from that seen in the periphery (see below) in that IL-1 production was not involved as the endogenous IL-1 receptor antagonist, IL-1ra, had no effect on the IL-6-induced hyperalgesia.[5] Injections of IL-1β

Cytokines in the Nervous System, edited by Nancy J. Rothwell. © 1996 R.G. Landes Company.

into the paraventricular nucleus of the hypothalamus have also been shown to produce a significant, albeit moderate, degree of thermal hyperalgesia.[6] More indirectly, recent work in our laboratory has shown that LPS administered icv induces a rapid thermal and mechanical hyperalgesia[7] and this is a central, not peripheral, effect of LPS. Although the involvement of cytokines was not addressed in that study it is well established that LPS is a potent stimulator of cytokine production.

Conversely, there are studies suggesting an analgesic effect of cytokines within the CNS and some of these demonstrate both pro- and anti-hyperalgesic effects within the same experimental paradigm. IL-1 and TNF-α have been shown to be anti-nociceptive in rats following icv administration, increasing withdrawal latencies to a noxious thermal stimulus.[8-10] Sellami and de Beaurepaire[6] showed that IL-1β was strongly analgesic when injected into thalamic nuclei in contrast to a hyperalgesic effect in the hypothalamus. These studies highlight some of the difficulties in interpreting behavioral studies using central administration of cytokines. For example, the increase in thermal nociceptive thresholds observed by Sarcadote et al[10] was limited to a short time period following icv administration and both IL-1α and TNF-α exhibited a bell-shaped dose response curve with the anti-nociceptive effect being lost at high doses (greater than 1 ng for TNF-α and 5-10 ng for IL-1). More recently, Oka et al[11] demonstrated differential effects of IL-1β on thermal hyperalgesia in rats depending on which hypothalamic nuclei were injected with the cytokine. IL-1β injected into the medial or lateral preoptic nuclei or the diagonal band of Broca was hyperalgesic but similar injections into the ventromedial nucleus proved to be analgesic. Finally, some studies have shown a lack of effect of IL-1α and IL-1β on nociceptive thresholds when given icv in rats with no significant analgesic effect by themselves nor any potentiation of morphine-induced analgesia.[12]

LOCALIZATION OF CYTOKINE RECEPTORS WITHIN THE CNS

If there is a central role for cytokines in pain then it would be expected that receptors for them will be present in areas associated with nociceptive processing. Although IL-1 receptors have been extensively described within the hypothalamic-pituitary axis[13] there are reports of more widespread localization. IL-1β and IL-1α receptors are particularly dense in the hippocampal region in mouse and rat[14-16] but IL-1 receptors have also been found to be more generally located throughout the brain with moderate to high levels of ^{125}I-IL-1 binding in cortical areas, thalamic and various brainstem nuclei areas known to be involved in nociceptive processing—and this binding was likely to be on neurones.[14] More recently, in situ hybridization techniques have demonstrated the expression of mRNA for IL-1 receptor mRNA in sub-cortical nuclei such as the thalamus and trigeminal nucleus.[17,18]

Cytokine receptors are located both on neuronal and non-neuronal cells. [125]IL-1 binding has been demonstrated on neurones in cortical and sub-cortical areas[14] and on astrocytes, but not microglial cells.[19]

In addition, IL-1 receptor mRNA was widely distributed throughout the CNS associated with endothelial cells of postcapillary venules.[17]

PRODUCTION OF CYTOKINES WITHIN THE CNS

The source of cytokines within the CNS is still not entirely clear but may occur, principally, by two mechanisms. They may enter the brain from the peripheral circulation, either via areas where the blood brain barrier is absent (the circumventricular organs) into the CSF and thence throughout the ventricular systems, or they may enter by an active transport mechanism. Interestingly, one study has suggested that the majority of this entry is into cortical areas[20] although it has been questioned whether this is a significant mechanism in terms of absolute levels of cytokines.[2,3]

The other major source of cytokines within the CNS is synthesis and release of them by the neuropil itself. With respect to nociception, the possibility of cytokine production within areas associated with the processing of nociceptive information is of particular relevance and there is good evidence that cytokines can be synthesized and released from both neuronal and non-neuronal cells in the CNS.

There is evidence that neurones themselves can synthesise cytokines. On the basis of immunohistochemical techniques, IL-1α and IL-1β have been identified in neurones and fibers within, predominantly, the hypothalamic nuclei and the hippocampus.[21,22] Immunoreactive IL-1β has also been identified in post mortem human brain.[23] In situ hybridization studies, however, have identified mRNA for IL-1β in more widespread areas including thalamic nuclei and trigeminal nuclei which are known to be associated with nociceptive processing.[18,21,24,25]

The other major possible site of production of cytokines within the CNS are immunocompetent cells. It is well established that in the periphery such cells, particularly monocytes/macrophages, are a major source of virtually all cytokines including interleukins, interferon and TNF-α.[26] There are several reports that immunocompetent cells within the CNS can release IL-1β when activated by trauma or inflammation. Amoeboid microglia and astrocytes have been shown to release IL-1β and TNF-α in vitro[27,28] and expression of mRNA for IL-1β and TNF-α has been demonstrated in astrocytes and microglia.[29,30]

ACTIVATION OF CNS NEURONES BY CYTOKINES

If cytokines are involved in nociceptive processing at the central level then it follows that they must be able to modulate the activity of those neurones involved in these processes, either directly or indirectly. Although there has not been a great deal of work in this area, activation of neurones by cytokines have been demonstrated in several ways.

Following peripheral application of IL-1 there is expression of mRNA for the immediate-early gene c-fos in the hypothalamic area and the parts of the limbic system within 1 hour. However, at later time points there was expression of c-fos more widely at the margins of the brain, suggesting more widespread activation of neurones.[31]

Using electrophysiological techniques, recombinant human IL-1β injected into the lateral ventricle enhanced the responses of wide dynamic neurones in the trigeminal nucleus to noxious stimulation and this was blocked by pretreatment with IL-1ra.[32] This action of IL-1β appears to be specific for nociceptive-processing neurones as the activity of those that responded to innocuous brushing in their peripheral field was not affected.[32]

It has also been shown that another cytokine, NGF, can excite cholinergic neurones in septal brain grafts *in oculo* when applied by iontophoresis.[33] An earlier study had shown that cholinergic neurones in the forebrain respond to NGF with an increase in activity of choline acetyltransferase[34] although there is, of course, less of an obvious correlation between these effects and nociception.

It is not clear whether this activation of neurones by cytokines is a direct effect or indirect but it is likely that prostanoids are involved. IL-1β given intravenously induces an increase in PGE2 levels in hypothalamic nuclei as well as more generally within the CSF and IL-1β-induced ACTH release is blocked by indomethacin.[35,36] IL-1β has also been shown to increase PGE2 production in rat astrocytes in vitro.[37] More specifically with respect to pain, thermal hyperalgesia induced by central administration of IL-1β and IL-6 is prevented by inhibitors of prostanoid synthesis.[5,32] IL-1β-induced enhancement of responses of neurones in the trigeminal nucleus to noxious stimulation is prevented by Na-salicylate.[32]

The anti-nociceptive actions of cytokines such as TNF-α and IL-1α may be via activation of central nor-adrenergic systems and/or CRF production.[38]

CYTOKINES AND THE PERIPHERAL NERVOUS SYSTEM

Although there is a substantial body of literature relating to cytokines, particularly NGF within the spinal cord and sensory ganglia, this is mainly concerned with nerve injury and regeneration with very few studies directly related to nociception. Most of the cytokines have been demonstrated in the spinal cord and sensory ganglia but any role with respect to nociceptive processing remains largely unclear.

Outside of the spinal cord, there is much more evidence for a role for cytokines in nociceptive processing. There is a substantial body of literature relating to the inflammatory properties of cytokines but this section will focus on those studies relating to the nociceptive actions of cytokines.

INTERLEUKIN-1

This cytokine has been studied most extensively and has been shown to have nociceptive actions both in vivo and in vitro models of pain and hyperalgesia.

Behavioral studies

One of the first studies to demonstrate behavioral hyperalgesia with cytokines showed that intraplantar injections of IL-1β in the rat induced mechanical hyperalgesia.[39] Subsequently, there have been several reports of IL-1β-induced mechanical and thermal hyperalgesia.[4,41-44] This action of IL-1β is specific to its receptor as its hyperalgesic actions are antagonized by antisera to IL-1β, IL-1ra and a tripeptide antagonist of IL-1β, Lys-D-Pro-Thr[39-42] (Fig. 10.1).

IL-1β is particularly potent in inducing behavioral hyperalgesia. Intraperitoneal injections of IL-1β have been shown to induce mechanical and behavioral hyperalgesia in the rat in the μg/kg range[4,41,45] and with local administration of IL-1β pg amounts are sufficient to induce long lasting mechanical and thermal hyperalgesia.[39,42,43,46] The potency of IL-1β as a hyperalgesic agent has been demonstrated in reports showing that a single intraplantar injection of IL-1β in the rat induces mechanical and thermal hyperalgesia not only in the injected paw but also in the contralateral paw.[39,43] The study by Ferriera et al[39] also suggested that the contralateral effect was a systemic action of IL-1β since bilateral hyperalgesia was also seen after intraperitoneal administration.

In contrast to these studies, Follenfant et al[41] saw a bilateral hyperalgesia after intraperitoneal administration of IL-1β in rats, but could only demonstrate an ipsilateral effect after local administration into the paw. Furthermore, this study failed to show a thermal hyperalgesia with IL-1β in mice. There is no clear explanation at present for these discrepancies, particularly as the same form of IL-1β was used. However, since IL-1β is produced and acts upon a variety of cell types, including macrophages, fibroblasts, neutrophils, lymphocytes, epithelial cells, endothelial cells and hepatocytes, the action of exogenously administered IL-1β may vary depending upon the cellular environment. Furthermore, the observed effects may vary depending upon dose of IL-1β used and the timepoint studied. However, what is clear from the above studies is that IL-1β is remarkably potent in inducing behavioral hyperalgesia.

Mechanisms of action of IL-1β

Studies addressing the mechanism of action of IL-1β in inducing behavioral hyperalgesia have used both systemic and local administration of IL-1β.

A series of studies by Watkins and colleagues[4,44,47] have looked in detail at the possible mechanisms of hyperalgesia after systemic

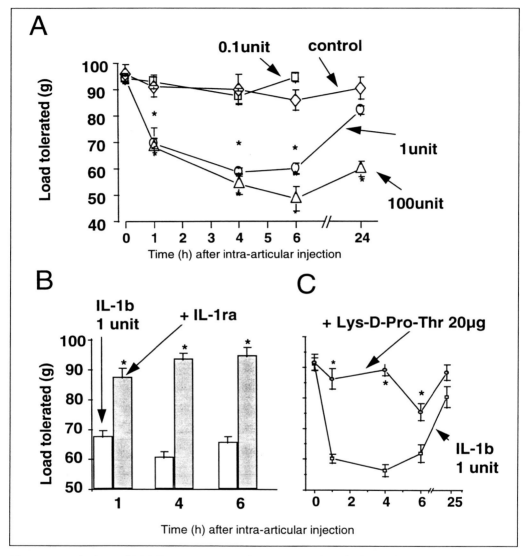

Fig. 10.1. Load tolerated (g) following intra-articular injections of interleukin 1β (IL-1β) into the knee joint of a rat. A. shows the mechanical hyperalgesia that develops for up to 24 h following intra-articular injections of 0.1, 1 and 100 units of IL-1β. Control animals received vehicle alone.

In B interleukin-1 receptor antagonist (IL-1ra 0.1 μg) was co-administered with IL-1β (shaded bars). The hyperalgesia induced by IL-1β (open bars) was reversed at all time points studied.

In C the IL-1β antagonist, Lys-D-Pro-Thr was co-administered with IL-1β and, similarly, prevented the development of mechanical hyperalgesia.

All values are mean ±s.e.m., n = 8-10/group. In A and C: *p < 0.05 ANOVA followed by post hoc analysis of means; in B *p < 0.5 students t-test.

administration of cytokines. Such studies have shown that IL-1β can induce thermal hyperalgesia after intraperitoneal administration, characterised by a reduction in withdrawal latencies in a well-characterized test of thermal nociception, the tail flick test.[45] The studies by Watkins and colleagues[4,44,47] are particularly interesting not only because they demonstrated IL-1β-induced hyperalgesia but also because they provided evidence that activation of vagal afferents was the predominant neural pathway for this hyperalgesia.[4] The same group had also shown that LPS-induced hyperalgesia was similarly blocked by vagotomy and by pretreatment with gandolium chloride.[4,47] As gandolium chloride inhibits the action of macrophages specific to the liver (Kupffer cells) and macrophages are well established to be a major source of cytokines, this suggests that one mechanism of cytokine-induced hyperalgesia in the periphery is via activation of hepatic vagal afferents by cytokines released from resident macrophages. Such studies demonstrate that increased peripheral levels of cytokines can cause hyperalgesia at distant peripheral sites via neural mechanisms involving central sites.

Experiments involving local administration of cytokines have also supported a role for IL-1β in the mediation of inflammatory pain. IL-1β causes hyperalgesia when injected locally into either the rat paw[39,41] or knee joint.[42] However, it is still not clear whether IL-1β acts directly on the nociceptive neurone to cause sensitization or activation, or whether its actions are indirect via release of secondary mediators such as kinins or prostaglandins.

This local IL-1β-induced hyperalgesia does not appear to involve the production of other cytokines, at least in the rat paw, as antisera to IL-8, IL-6 and TNF-α do not reduce the hyperalgesia induced by IL-1β[40] and neither does it appear to involve the sympathetic nervous system.[40,48]

IL-1β has been shown to cause an increased firing of fibers when injected into the hind paw or the knee joint of the rat.[49,50] In the paw not only was spontaneous firing of sensory nerves increased by IL-1β but also neuronal responses to noxious cold and heat stimuli were enhanced by IL-1β.[49] In addition to noxious thermal stimuli IL-1β also decreased the threshold for firing to pressure stimulation.[49] More recently, it has been shown that joint mechanonociceptors in vitro respond to IL-1β with a sustained increase in activity.[50] Although in these studies the effects of IL-1β were apparent within a timescale of up to 10 minutes, it is still not possible to conclude whether IL-1β was having a direct effect on nociceptive neurones or acting indirectly via release of other sensitizing agents such as kinins or prostaglandins. However, in acutely dissociated dorsal root ganglion cells of the rat, IL-1β causes an immediate increase in intracellular concentration of Ca^{2+} ions, suggesting that IL-1β may be acting directly on the sensory neurones.[51]

INTERLEUKIN-1 AND PROSTAGLANDINS

There is no clear consensus with respect to the involvement of prostaglandins in the behavioral hyperalgesia seen with IL-1β. Indomethacin has been shown to block IL-1β responses in some studies[39,40,42] and to be ineffective in others.[41] However, in vitro studies have shown that IL-1β causes a large increase in the levels of PGE2 in human synovial cells. In these cells both cytosolic PLA_2 and cyclooxygenase-2 (COX-2) mRNA were upregulated upon 5 h exposure to IL-1 whereas the levels of constitutive COX-1 were not changed.[52,53] IL-1 has also been proposed as a trigger for the induction of COX-2 by inducing the expression of mRNA for COX-2 as well as increasing its activity in many cell types and cell.[54,55,56]

INTERLEUKIN-1 AND KININS

There is a substantial body of evidence in the literature suggesting a complex interaction between IL-1β, cytokines and the kinin system; the evidence as it relates to nociception is discussed later.

ENDOGENOUS INTERLEUKIN-1 ANTAGONISTS

Production of IL-1 is tightly regulated within the body in a complex manner. In addition to the endogenous IL-1 receptor antagonist, IL-1ra the cytokine system has several other ways of limiting the action of IL-1.[57] There are two distinct types of IL-1 receptors, namely, the IL-1 RI and IL-1 RII receptors; the IL-1 RI receptor being present mainly on T-cells, endothelial and epithelial cells and chrondrocytes, whereas the IL-1 RII is present on B cells, neutrophils and macrophages, (see Dayer and Burger[58] for review). The type II IL-1 receptor has been referred to as a "decoy" receptor, since the binding of IL-1 to it is not translated to any signal transduction.[59,60] Also the extracellular domains of both types of IL-1 receptors can be shed and act as soluble receptors. These soluble receptors may play important regulatory roles, since they still bind IL-1. As IL-1 binds to the type II soluble receptor with a greater affinity than IL-1ra, this receptor could further act to reduce the biological actions of IL-1.[58]

There is recent evidence for even more complex regulatory mechanisms limiting the hyperalgesic actions of IL-1β. Stein and colleagues[61] have shown that subsequent to the establishment of an inflammatory condition in the rat paw, administration of IL-1β can be analgesic due to release of peripheral opioid peptides.[61] This may seem to be contradictory to the reports showing peripheral algesic actions of IL-1β, as discussed above, however this anti-hyperalgesic action of IL-1β may be related to the local concentrations of IL-1β present in these studies. As levels of IL-1β would already be elevated in the inflamed paw[62] it maybe that the anti-hyperalgesic effect of additional IL-1β could be an indication of a physiological control mechanism to limit IL-1β-induced hyperalgesia via the release of endogenous opioids.

In general, the majority of studies have shown that exogenously injected IL-1β is pro-nociceptive in animals but, do not clearly address the question as to whether IL-1β has a role in pain perception in man. This question will only be answered definitively if and when an IL-1 antagonist is used clinically but the animal studies do suggest that IL-1 may be an important pain mediator in humans. At present there are few clinical studies relating IL-1β, or any other cytokine with pain. Specifically pain although a recent report has shown IL-1α, IL-1β, IL-6 levels to be increased in herniated disks in man and that IL-1α increased PGE2 production from excised herniated disk tissue.[63]

INTERLEUKIN-2, -6 AND -8

There is little experimental evidence directly linking these cytokines with nociception but some recent studies from behavioral paradigms in rats have suggested a role for IL-2, IL-6 and IL-8 in inflammatory hyperalgesia.

IL-2 has been shown to cause behavioral hyperalgesia when injected locally into the knee joint of the rat.[42] IL-8 causes hyperalgesia when injected locally into either the paw or the knee joint and IL-6 causes hyperalgesia when injected into the paw but not the knee joint of rats.[40,42,48]

The experimental evidence from our studies using intra-articular injections of IL-2 suggests that IL-2-induced hyperalgesia in the knee joint involves endogenous production of IL-1, since the IL-2-induced hyperalgesia was blocked by IL-1ra.[42] The involvement of IL-1 in IL-2 induced hyperalgesia is not suprising since IL-2 has been shown to induce the production of a variety of cytokines including IL-1, TNF-α and IL-8, thus its effects could be entirely due to the production of secondary cytokines.[64-67] However, IL-2 has also been shown to activate a sub-population of cutaneous nociceptors in rat skin, although antagonists were not used in this study so it is not possible to conclude whether IL-2 was acting directly or via production of secondary mediators.[68]

IL-8 has been most extensively studied with respect to its chemotactic effects but it has been shown to produce behavioral hyperalgesia.[40,42,48] There is a difference, however, between these studies as to the degree of involvement of IL-1β. In the knee joint IL-8-induced hyperalgesia was antagonized by IL-1ra, whereas in the paw antibodies to IL-1 had no effect. In addition, an involvement of the sympathetic system in its actions has been suggested.[40,48]

TUMOR NECROSIS FACTOR

As with IL-1β-induced hyperalgesia, most of the experimental evidence linking TNF-α in hyperalgesia are from behavioral models of hyperalgesia. The biological properties of TNF-α are very similar to those of IL-1β,[69] and thus studies into TNF-induced hyperalgesia are often compared to those of IL-1. Behavioral hyperalgesia has been demonstrated after local administration of TNF-α into rat paws and following systemic administration using a tail flick hyperalgesia model.[42,44] As with the action of IL-1, there were distinct differences when comparing local actions with the ip effects of TNF-α. When administered locally TNF-α, like IL-1β, induced an indomethacin sensitive hyperalgesia, which was also antagonized by antisera to IL-1β, and the tripeptide antagonist of IL-1β Lys-D-Pro-Thr.[40] The hyperalgesia was only blocked by about 50% by IL-1 antagonists or indomethacin, however, there seemed to be a sympathetic component to TNF-induced hyperalgesia.[40] Thus the local effects of TNF seem to involve some endogenous production of IL-1. The converse was not true because IL-1-induced hyperalgesia was refractory to TNF antisera.

When TNF-α was administered ip a similar hyperalgesia was seen in tail flick experiments to that produced by IL-1β. The hyperalgesia was antagonized by subdiaphragmatic vagotomy, suggesting a neural mode of action. In addition the TNF-α-induced hyperalgesia was antagonized by IL-1ra, thus it was probably acting via endogenous production of IL-1β in this model.[44]

Although the experimental evidence linking TNF-α as a possible pain mediator in animals is less than for IL-1β, recent human experiments show that blocking TNF may have beneficial effects.[70] CDP571, an engineered antibody which neutralizes human TNF-α, caused a reduction in both pain and arthritis symptoms. However, since the antibody treatment caused a general reduction in all signs of arthritis, including joint swelling, the reduction in pain could have been due to the anti-inflammatory effects of the drug reducing the presence of other mediators of hyperalgesia in the joint.

TNF-α has been shown, to date, to act via two distinct receptors TNF-RI and TNF-RII, and, as with the IL-1 system, the two receptors are capable of shedding their extracellular domains which can then act as specific TNF-α inhibitors. Since the binding affinity of TNF-α for the membrane bound and soluble receptors is similar, large amounts of receptor are required to block the action of TNF-α, thus soluble receptors have been considered unsuitable for therapeutic development.[58,71]

NERVE GROWTH FACTOR

Studies in neonatal and adult rats have shown that the NGF takes on a different physiological role in mature animals.[72,73] Adult sensory neurones possess functional high affinity NGF receptors[74] and it has been proposed that NGF may be the linchpin between inflammation and hyperalgesia.[73]

Several groups have looked at the behavioral hyperalgesia after peripheral administration of NGF. After ip injection of NGF[75] thermal and mechanical thresholds to paw withdrawal latency and threshold were reduced, which is indicative of hyperalgesia. The timecourse of thermal and mechanical hyperalgesia differed, however, with thermal hyperalgesia apparent within minutes and the mechanical response delayed several hours. The initial thermal responses (10 min-3 h post injection of NGF) were not apparent in animals pre-treated with compound 48/80 to degranulate mast cells, whereas by 7 h post administration the thermal hyperalgesia was at the same level as control animals. The 5-HT receptor antagonists methiothepin and ICS 205-930 also antagonized the initial thermal hyperalgesia with NGF, thus suggesting that NGF may, initially, act via release of 5-HT from mast cells. The mechanical and late thermal hyperalgesia appeared to involve centrally mediated mechanisms since the NMDA receptor antagonist MK 801 could block these effects.[75]

The mechanism of hyperalgesia may be different in studies looking at locally mediated effects of NGF. After local administration of NGF into the paws of rats a marked thermal hyperalgesia was apparent from 15 min and lasted for 24 h at the highest dose tested (500 ng/paw). This hyperalgesia was reduced in animals which had undergone either chemical or surgical sympathectomy prior to NGF treatment. However, the initial (60 min) phase of the hyperalgesia was not affected by the sympathectomy treatment.[76] It is difficult to draw general conclusions at present from these studies using low dose local administration of NGF and those employing systemic administration of NGF. For example, the possible role of mast cells has not been explored with respect to local administration. Conversely, studies using systemic administration did not investigate a possible sympathetic component to their hyperalgesia.

A recent study looking at the involvement of NGF in hyperalgesia[62] has looked more closely at the possible link between local NGF in the periphery and inflammatory hyperalgesia. This study showed that after a peripheral inflammatory insult with CFA injected directly into the hind paw, local levels of NGF increased and the resultant hyperalgesia could be attenuated with an antibody to NGF. Of particular interest in this study was the observation that peripherally administered IL-1β increased NGF levels and this increase was blocked by IL-1ra, however, local NGF administration did not increase IL-1β levels. Thus these observations suggest that in inflammatory hyperalgesia IL-1β acts to produce NGF, which then acts to cause hyperalgesia either by directly acting on NGF receptors on sensory neurones or indirectly via release of another peripheral sensitizing agent. Alternately, however, since NGF produces minimal inflammation, NGF could be acting via IL-1 production, but the levels produced could be below the assay detection limits of this study. In other behavioral studies injection of 1-10 pg of IL-1β can cause mechanical hyperalgesia,[39,42] and it would not be unreasonable to expect endogenous levels to be below this. In view of the many ways that the effectiveness of endogenous IL-1 can be limited, as discussed above (e.g., IL-1ra, 'decoy' receptors etc), it is difficult to correlate increased levels of IL-1 with a specific nociceptive response.

'INHIBITORY' CYTOKINES

Recently a group of cytokines have been identified which appear to have an 'anti-inflammatory' role and can be considered to be 'inhibitory' cytokines. The best characterised of these 'inhibitory' cytokines are IL-4, IL-10 and IL-13. IL-13 has very similar actions to IL-4 which may be related to the fact that although they have distinct receptors (IL-4R and IL-13R) the receptors share common subunits.[77]

Both IL-10 and IL-4 can inhibit the prostanoid synthesis of LPS-stimulated monocytes and this appears to be due to suppression of the gene expression of the inducible isoform of cyclooxygenase (COX-2). IL-10 acts to both accelerate the degradation of COX-2 mRNA and inhibits LPS-induced COX-2 gene transcription.[78] This action on COX-2 production may be of particular relevance to inflammatory pain since COX-2 is virtually absent in normal tissues but is induced following an inflammatory insult such as carrageenin or endotoxin administration.[54,79-81]

Both IL-4 and IL-10 have been shown to reduce the spontaneous production of IL-1β in cultured synovium pieces from arthritic patients.[82] IL-4 also suppresses the expression of IL-1 in LPS stimulated human monocytes and decreased IL-1 mRNA levels by repressing IL-1 gene transcription.[83] IL-10 has been shown to inhibit nuclear localisation of NFκB, a transcription factor involved in the expression of inflammatory cytokine genes.[84] In human synovial cells cultured from rheumatoid arthritis patients, IL-4 inhibits the production of PGE2 induced by both IL-1β and TNF-α.[85]

With respect to nociception, however, there is, to the best of our knowledge, only one study to date assessing the antinociceptive actions of "inhibitory cytokines" in behavioral models of inflammatory hyperalgesia.[86] In this study, which used the rat paw pressure model, IL-10 was shown to inhibit the hyperalgesia produced by BK, IL-1β, TNF-a or IL-6. However it was ineffective against the hyperalgesia caused by IL-8 and PGE2. Such an action of IL-10 would be consistent with its anti-inflammatory properties and Poole and colleagues[86] present evidence for suppression of COX-2 being one of the underlying mechanisms for the anti-hyperalgesic action of IL-10. An endogenous role for IL-10 in inflammatory hyperalgesia was supported by the potentiation of carrageenin-induced hyperalgesia by a monoclonal antibody to IL-10.[86]

CYTOKINE-INDUCED HYPERALGESIA AND KININS: THE 'CYTOKININ AXIS'

Kinins are a family of small pro-inflammatory peptides of which the two most studied are bradykinin and kallidin, a nona- and decapeptide respectively. These kinins are produced locally, from precursor molecules, in response to tissue injury or trauma and, particularly, during inflammation. Kinins mediate their effects via two main classes of receptor, B_1 and B_2 with bradykinin (BK) and kallidin acting on the B_2 receptor.[87] The B_1 receptor is not normally expressed, but following an inflammatory insult B_1 receptor-mediated responses become apparent. The preferential agonists for the B_1 receptor are the only active metabolites of bradykinin or kallidin, desArg9BK or desArg10 kallidin, which lack the carboxy-terminal arginine see Bhoola et al[88] for a review.

Kinins exert a variety of effects on many body tissues but it is their actions on nociceptive neurones which are of relevance to pain. The most studied kinin in this regard is bradykinin which has been shown to both activate and sensitize nociceptive afferents leading to both overt pain and hyperalgesia. Both of these actions have been shown, so far, to be mediated primarily by the B2 receptor. Recently, however, several studies have showed that during inflammation there is an increasing contribution of B_1 receptors to the accompanying hyperalgesia.[89-92]

What relevance has this to any possible role of cytokines in peripheral hyperalgesia or pain?

There is a substantial body of work demonstrating a close interaction between kinins and cytokines. In particular, IL-1β upregulates B1 and B2 receptors and enhances responses to bradykinin and desArg9BK.[87,91-94] Conversely both bradykinin and desArg9BK have been shown to induce the release of IL-1 from macrophages.[95] As stated above, the expression of the B_1 appears to be dependent on the presence of an inflammatory reaction. More specifically, it has been shown both in vitro and in vivo that IL-1β, and other cytokines, are able to induce the B_1 receptor (see Marceau[96] for a review). With respect to pain and hyperalgesia, we have shown that, following the production of a local inflammatory response in the knee joint or hind paw of the rat, there is the induction of a B_1 receptor-mediated mechanical and thermal hyperalgesia.[91,92] Furthermore, we have demonstrated that local injections of IL-1β by itself is just as effective in inducing this B_1 receptor-hyperalgesia.[42,43] Interestingly, this hyperalgesia is not restricted to the site of injection of IL-1β but is also seen in the contralateral paw.[43] As previously mentioned, local administration of IL-1β itself has been shown to lead to 'distant' nociceptive responses and this phenomenon also appears to be true for the induction of kinin B_1 receptor-mediated responses (Fig. 10.2). At the level of the nociceptor it has recently been shown that, subsequent to treatment with IL-1β, joint mechanonociceptors in vitro respond to desArg9BK with an increase in firing rate as well as showing enhanced responses to BK.[50]

Further studies from our laboratory have suggested that the interaction between cytokines and kinins with respect to nociception is a complex one. The desArg9BK hyperalgesia that occurs following either a non-specific inflammatory insult or cytokine administration is reversed by IL-1ra, implying that production of IL-1β is necessary for the maintenance of this B_1 receptor-mediated nociceptive response.

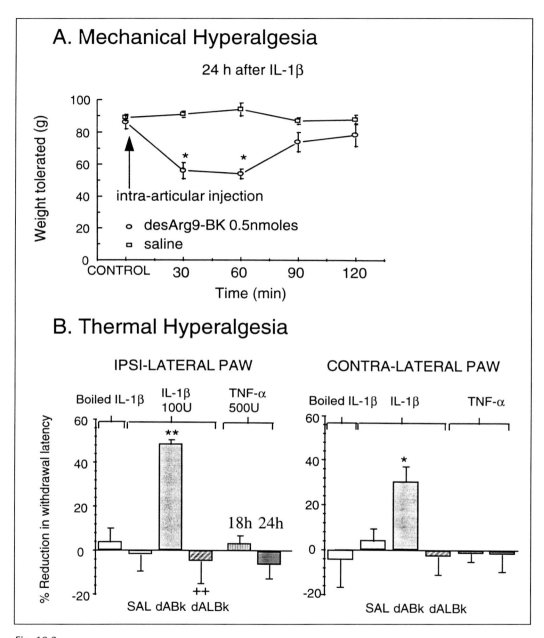

Fig. 10.2.

However, mechanical hyperalgesia induced by IL-1β itself is also blocked by both B_1 and B_2 kinin receptor antagonists, desArg9Leu8BK and icatibant, respectively.[42] This suggests that IL-1β-induced hyperalgesia is dependent itself on both kinin B_1 and B_2 kinin receptors (Fig. 10.3). In view of the inflammatory properties of IL-1β and other cytokines,[97,98] it would not be surprising that one consequence of cytokine production would be kinin production. Surprisingly, there are no studies known to us which directly address this, however, TNF-α induces the release of kininogen from murine fibroblasts[99] which could be converted to kinins in inflammatory exudates.

There are only a few reports linking other cytokines with kinins with respect to nociception, but one recent interesting study showed that desArg9Leu8BK reduced the late thermal, but not mechanical, hyperalgesia induced by systemic administration of NGF.[100] The B_2 receptor antagonist, icatibant, was without effect.

As with most behavioral studies it is extremely difficult to conclude at what level such bi-directional interactions occur. What these studies do suggest, though, is that there appears to be a powerful interdependency of kinins and cytokines, at least with respect to nociception and pain. This 'cytokinin axis' may be of great significance with respect to the regulation of nociceptive neurone sensitivity during an inflammatory episode. It is important to note, with respect to such a role for kinins that, unlike the cytokines, there is no need for activation of immunocompetent cells for the production of kinins to occur.

Fig. 10.2. (see opposite page) In A the weight tolerated (g) following intra-articular injections of desArg9BK (open circles) or saline (open squares) into the knee joint of a rat 24 h after an intra-articular injection of interleukin-1β (1 unit). At this time point the hyperalgesia induced by IL-1β had fully recovered to normal values. desArg9BK is without effect in normal untreated knees (data not shown) but following pretreatment with IL-1β there is a reduction in tolerated load indicative of mechanical hyperalgesia.
In B, the percentage reduction in paw withdrawal latency to a noxious thermal heat beam, compared with pre-drug measurement in the same animal are shown for the ipsilateral paw and contralateral paw following intraplantar injections into the ipsilateral paw of, boiled interleukin-1β (IL-1β), IL-1β 100 units (U) and tumour necrosis factor-α (TNF-α, 500 U) with subsequent systemic injection of saline or desArg9BK (10 nmole/kg iv). All measurements were made 48 hrs after cytokine administration except in the case of TNF-α when an additional time point of 18 h was included. For both paws the open bars are boiled IL-1β followed by desArg9BK. The three bars grouped under IL-1b represent pretreatment with IL-1β followed by saline (SAL), desArg9BK (dABk,) and desArg9BK together with desArg9Leu8BK (dALBk). The two bars grouped under TNF-α refer to injections of desArg9BK 48h and 18h after TNF-α administration. DesArg9BK-induced thermal hyperalgesia was only observed following intraplantar IL-1β and was present in both the ispilateral and contralateral paws. There was no hyperalgesia when desArg9BK was injected subsequent to boiled IL-1β or TNF-α, at either time point, in either paw. Co-administration of desArg9Leu8BK (200 nmole/kg iv) completely abolished the desArg9BK-induced thermal hyperalgesia, when present, in both paws. All values are mean ± s.e.m., n = 4, *p = < 0.01, **p = < 0.001, paired t-test.

Kinin production can be triggered by the simple presence of endotoxin or damaged cell surfaces. They are therfore, likely to be the first
pro-inflammatory mediators to be formed following trauma or inflammation and thus play a key role in regulating the actions of later mediators, such as cytokines, with respect to nociception.

SUMMARY AND CONCLUSIONS

Cytokines clearly have many roles within the body. In contrast to
their clearly established roles in neuroendocrine function, inflammation, pyrogenesis and regulation of nerve growth and differentiation,
their involvement in the transduction and modulation of nociception
is still poorly understood. It is clear that some of them, particularly
IL-1β and TNF-α, do have a powerful modulatory effect on nociception
but many questions remained unanswered or unaddressed.

Some of the critical questions are do the cytokines act directly on
nociceptive neurones or is their action indirect and, if a direct action,
what transduction mechanisms are involved with respect to nociception?

At present, the studies relating to cytokines and nociception have
concentrated on inflammatory hyperalgesia but is there a role for this
class of mediators in other types of pain, for example, neuropathic
pain?

There is clearly the potential for both a peripheral and central
role for cytokines in nociceptive processing but the precise mechanisms
involved remain to be clarified. In addition, there is increasing evidence that cytokines and other inflammatory mediators, particularly
the kinins, have a close and complex interdependency with respect to
the initiation and maintenance of inflammatory hyperalgesic in the
periphery and, possibly, with the CNS as well. There is evidence both
that cytokines are the initiators of a sequence of production of inflammatory mediators leading to hyperalgesia and the reverse, that kinins or other early inflammatory mediators, may be the 'trigger' for
cytokine production. The precise sequence of events may well be different depending on the inflammatory stimulus and the tissues involved.

*Fig. 10.3. (see opposite page) A and B show the load tolerated (g) following intra-articular
injections of desArg9BK into the knee joint of rats. In A the open bar shows the
hyperalgesia induced by desArg9BK in knee joints pretreated 24 h before with interleukin-
1β (IL-1β) in order to induce a B₁ receptor mediated response (compared to those animals
which only received IL-1β; the 'control' bar). This hyperalgesic response to desArg9BK is
prevented by co-administration of interleukin-1 receptor antagonist or by the tripeptide
antagonist of IL-1, Lys-D-Pro-Thr (KDPT) with desArg9BK.
B shows that IL-1β-induced hyperalgesia is also reversed by the administration of a B₁ and
a B₂ receptor antagonist, desArg9Leu8Bk and HOE140, respectively. desArg9Leu8Bk and
HOE140 were administered iv 4h after intra-articular injection of IL-1β (0.1 unit) when the
IL-1β-induced hyperalgesia was maximal.
All values are mean ± s.e.m., n = 6-8/group, *p < 0.05 ANOVA followed by post hoc
analysis of means.*

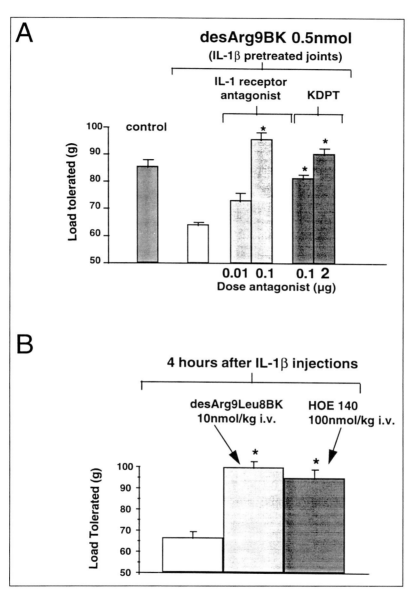

Fig. 10.3.

Not only is this a new and interesting area for further research but it is possible that it may lead to a therapeutic application in the future. It is not surprising that the cytokine system is currently being targeted for the development of novel anti-inflammatory drugs, but it is also possible that cytokines may be fruitful as a target for novel analgesic drugs as well.

REFERENCES

1. Rothwell NJ. CNS regulation of thermogenesis. Crit Rev Neurobiol 1994; 8:1-10.
2. Watkins LR, Maier SF, Goehler LE. Immune activation: the role of pro-inflammatory cytokines in inflammation, illness responses and pathological pain states. Pain 1995; 63:289-302.
3. Watkins LR, Maier SF, Goehler LE. Cytokine-to-brain communication: a review and analysis of alternative mechanisms. Life Sci 1995; 57(11):1011-26.
4. Watkins LR, Wiertelak EP, Goehler Le et al. Characterization of cytokine-induced hyperalgesia. Brain Res 1994; 654:15-26.
5. Oka T, Oka K, Hosoi M et al. Intracerebroventricular injection of interleukin-6 induces thermal hyperalgesia in rats. Brain Res 1995; 692(1-2):123-8.
6. Sellami S, de Beaurepaire R. Hypothalamic and thalamic sites of action of interleukin-1 beta on food intake, body temperature and pain sensitivity in the rat. Brain Res 1995; 694:69-77.
7. Walker K, Dray A, Perkins M. Hyperalgesia in rats following intracerebroventricular administration of endotoxin: Effect of bradykinin B_1 and B_2 receptor antagonist treatment. Pain 1996; In Press.
8. Bianchi M, Sacerdote P, Ricciardi Castagnoli P et al. Central effects of tumor necrosis factor alpha and interleukin-1 alpha on nociceptive thresholds and spontaneous locomotor activity. Neurosci Lett 1992; 148:76-80.
9. Bianchi M, Panerai AE. CRH and the noradrenergic system mediate the antinociceptive effect of central interleukin-1 alpha in the rat. Brain Res Bull 1995; 36:113-7.
10. Sacerdote P, Bianchi M, Ricciardi Castagnoli P et al. Tumor necrosis factor alpha and interleukin-1 alpha increase pain thresholds in the rat. Ann N Y Acad Sci 1992; 650:197-201.
11. Oka T, Oka K, Hosoi M et al. The opposing effects of interleukin-1 beta microinjected into the preoptic hypothalamus and the ventromedial hypothalamus on nociceptive behavior in rats. Brain Res 1995; 700(1-2):271-8.
12. Adams JU, Bussiere JL, Geller EB, Adler MW. Pyrogenic doses of intracerebroventricular interleukin-1 did not induce analgesia in the rat hot-plate or cold-water tail-flick tests. Life Sci 1993; 53:1401-9.
13. Haour F, Ban C, Marquette G et al. Brain interleukin-1 receptors: mapping, characterization and modulation. In: Rothwell NJ, Dantzer RD eds. Interleukin-1 in the Brain. Oxford: Pergamon Press, 1992; 13-25.

14. Farrar WL, Kilian PL, Ruff MR et al. Visualization and characterization of interleukin 1 receptors in brain. J Immunol 1987; 139:459-63.

15. Takao T, Tracey DE, Mitchell WM et al. Interleukin-1 receptors in mouse brain: characterization and neuronal localization. Endocrinology 1990; 127:3070-8.

16. Haour FG, Ban EM, Milon GM et al. Brain interleukin-1 receptors: characterization and modulation after lipopolysaccharide injection. Progress in NeuroEndocrinImmunology 1990; 3:196-204.

17. Cunningham ET Jr, Wada E, Carter DB et al. In situ histochemical localization of type I interleukin-1 receptor messenger RNA in the central nervous system, pituitary, and adrenal gland of the mouse. J Neurosci 1992; 12:1101-14.

18. Yabuuchi K, Minami M, Katsumata S et al. Localization of type I interleukin-1 receptor mRNA in the rat brain. Brain Res Mol Brain Res 1994; 27:27-36.

19. Ban EM, Sarlieve LL, Haour FG. Interleukin-1 binding sites on astrocytes. Neuroscience 1993; 52:725-33.

20. Banks WA, Kastin AJ, Durham DA. Bidirectional transport of interleukin-1 alpha across the blood-brain barrier. Brain Res Bull 1989; 23:433-7.

21. Lechan RM, Toni R, Clark BD et al. Immunoreactive interleukin-1 beta localization in the rat forebrain. Brain Res 1990; 514:135-40.

22. Schultzberg M. Location of interleukin-1 in the nervous system. In: Rothwell N, Dantzer RD, eds. Interleukin-1 in the Brain. Oxford: Pergamon Press, 1992; 1-11.

23. Breder CD, Dinarello CA, Saper CB. Interleukin-1 immunoreactive innervation of the human hypothalamus. Science 1988; 240:321-4.

24. Bandtlow CE, Meyer M, Lindholm D, Spranger M, Heumann R, Thoenen H. Regional and cellular codistribution of interleukin 1 beta and nerve growth factor mRNA in the adult rat brain: possible relationship to the regulation of nerve growth factor synthesis. J Cell Biol 1990; 111:1701-11.

25. Yabuuchi K, Minami M, Katsumata S et al. In situ hybridization study of interleukin-1 beta mRNA induced by kainic acid in the rat brain. Brain Res Mol Brain Res 1993; 20:153-61.

26. Cavaillon JM. Cytokines and macrophages. Biomed Pharmacother 1994; 48(10):445-53.

27. Giulian D, Baker TJ, Shih LC et al. Interleukin 1 of the central nervous system is produced by ameboid microglia. J Exp Med 1986; 164:594-604.

28. Fontana A, Kristensen F, Dubs R, et al. Production of prostaglandin E and an interleukin-1 like factor by cultured astrocytes and C6 glioma cells. J Immunol 1982; 129:2413-9.

29. Chung IY, Norris JG, Benveniste EN. Differential tumor necrosis factor alpha expression by astrocytes from experimental allergic encephalomyelitis-susceptible and - resistant rat strains. J Exp Med 1991; 173:801-11.

30. Van Dam AM, Brouns M, Louisse S et al. Appearance of interleukin-1 in macrophages and in ramified microglia in the brain of endotoxin-treated rats: a pathway for the induction of non-specific symptoms of sickness?

Brain Res 1992; 588:291-6.

31. Brady LS, Lynn AB, Herkenham M, Gottesfeld Z. Systemic interleukin-1 induces early and late patterns of c-fos mRNA expression in brain. J Neurosci 1994; 14:4951-64.

32. Oka T, Aou S, Hori T. Intracerebroventricular injection of interleukin-1 beta enhances nociceptive neuronal responses of the trigeminal nucleus caudalis in rats. Brain Res 1994; 656:236-44.

33. Palmer MR, Eriksdotter Nilsson M, Henschen A et al. Nerve growth factor-induced excitation of selected neurons in the brain which is blocked by a low-affinity receptor antibody. Exp Brain Res 1993; 93:226-30.

34. Ebendal T. NGF in CNS: experimental data and clinical implications. Prog Growth Factor Res 1989; 1:143-59.

35. Komaki G, Arimura A, Koves K. Effect of intravenous injection of IL-1 beta on PGE2 levels in several brain areas as determined by microdialysis. Am J Physiol 1992; 262:E246-51.

36. Katsuura G, Gottschall PE, Dahl RR et al. Adrenocorticotropin release induced by intracerebroventricular injection of recombinant human interleukin-1 in rats: possible involvement of prostaglandin. Endocrinology 1988; 122:1773-9.

37. Katsuura G, Gottschall PE, Dahl RR et al. Interleukin-1 beta increases prostaglandin E_2 in rat astrocyte cultures: modulatory effect of neuropeptides. Endocrinology 1989; 124:3125-7.

38. Bianchi M, Panerai AE. CRH and the noradrenergic system mediate the antinociceptive effect of central interleukin-1 alpha in the rat. Brain Res Bull 1995; 36:113-7.

39. Ferreira SH, Lorenzetti BB, Bristow AF et al. Interleukin-1beta as a potent hyperalgesic agent antagonized by a tripeptide analogue. Nature 1988; 334:698-700.

40. Cunha FQ, Poole S, Lorenzetti BB et al. The pivotal role of tumour necrosis factor alpha in the development of inflammatory hyperalgesia. Br J Pharmacol 1992; 107:660-4.

41. Follenfant RL, Nakamura Craig M, Henderson B et al. Inhibition by neuropeptides of interleukin-1 beta-induced, prostaglandin-independent hyperalgesia. Br J Pharmacol 1989; 98:41-3.

42. Davis AJ, Perkins MN. The involvement of bradykinin B1 and B2 receptor mechanisms in cytokine-induced mechanical hyperalgesia in the rat. Br J Pharmacol 1994; 113:63-8.

43. Perkins MN, Kelly D. Interleukin-1 beta induced-desArg9bradykinin-mediated thermal hyperalgesia in the rat. Neuropharmacology 1994; 33:657-60.

44. Watkins LR, Goehler LE, Relton J et al. Mechanisms of tumor necrosis factor-alpha (tnf-alpha) hyperalgesia. Brain Res 1995; 692:244-50.

45. Maier SF, Wiertelak EP, Martin D et al. Interleukin-1 mediates the behavioral hyperalgesia produced by lithium chloride and endotoxin. Brain Res 1993; 623:321-4.

46. Perkins MN, Kelly D, Davis AJ. Bradykinin B-1 and B-2 receptor mecha-

nisms and cytokine- induced hyperalgesia in the rat. Can J Physiol Pharmacol 1995; 73:832-6.47.

47. Watkins LR, Wiertelak EP, Goehler et al. Neurocircuitry of illness-induced hyperalgesia. Brain Res 1994; 639:283-99.

48. Cunha, FQ, Lorenzetti, BB, Poole S et al. Interleukin-8 as a mediator of sympathetic pain. Br J Pharmacol 1991; 104:765-767.

49. Fukuoka H, Kawatani M, Hisamitsu T et al. Cutaneous hyperalgesia induced by peripheral injection of interleukin-1 beta in the rat. Brain Res 1994; 657:133-140.

50. Kelly DC, Ashgar AUR, McQueen DS et al. Effects of bradykinin and desArg9-bradykinin on afferent neural discharge in interleukin-1β-treated knee joints. Br J Pharmacol 1996; 117:90P.

51. Kawatani M, Birder L. Interleukin-1 facilitates Ca2+ release in acutely dissociated dorsal root ganglion (DRG) cells of rat. Neurosci Abstr 1992; 18:691 (Abstract).

52. Bathon JM, Croghan JC, MacGlashan DW Jr., Proud D. Bradykinin is a potent and relatively selective stimulus for cytosolic calcium elevation in human synovial cells. J Immunol 1994; 153(6):2600-8.

53. Angel J, Audubert F, Bismuth G, Fournier C. IL-1b amplifies bradykinin-induced prostaglandin E2 production via a phospholipase D-Linked mechanism. J Immunol 1994; 152:5032-40.

54. Chanmugam P, Feng L, Liou S et al. Radicicol, a protein tyrosine kinase inhibitor, suppresses the expression of mitogen-inducible cyclooxygenase in macrophages stimulated with lipopolysaccharide and in experimental glomerulonephritis. J Biol Chem 1995; 270:5418-26.

55. Endo T, Ogushi F, Sone S, et al. Induction of cyclooxygenase-2 is responsible for interleukin-1 beta- dependent prostaglandin E_2 synthesis by human lung fibroblasts. Am J Respir Cell Mol Biol 1995; 12:358-65.

56. Mitchell JA, Belvisi MG, Akarasereenont P et al. Induction of cyclooxygenase-2 by cytokines in human pulmonary epithelial cells: regulation by dexamethasone. Br J Pharmacol 1994; 113:1008-14.

57. Arend WP. Interleukin-1 receptor antagonist. Advances in Immunology 1993; 54:161-7.

58. Dayer JM, Burger D. Interleukin-1, tumor necrosis factor and their specific inhibitors. Eur Cytokine Netw 1994; 5:563-71.

59. Colotta F, Re F, Muzio M et al. Interleukin-1 type II receptor: a decoy target for IL-1 that is regulated by IL-4. Science 1993; 261(5120):472-5.

60. Re F, Muzio M, De Rossi M et al. The type II "receptor" as a decoy target for interleukin-1 in polymorphonuclear leukocytes: Characterization of induction by dexamethasone and ligand binding properties of the released decoy receptor. J Exp Med 1994; 179:739-43.

61. Schafer M, Carter L, Stein C. Interleukin 1 beta and corticotropin-releasing factor inhibit pain by releasing opioids from immune cells in inflamed tissue. Proc Natl Acad Sci USA 1994; 91:4219-23.

62. Safieh-Garabedian B, Poole S, Allchorne A et al. Contribution of interleukin-1 beta to the inflammation- induced increase in nerve growth

factor levels and inflammatory hyperalgesia. Br J Pharmacol 1995; 115:1265-75.

63. Takahashi H. A mechanism for sciatic pain caused by lumbar disc herniation—involvement of inflammatory cytokines with sciatic pain. Nippon Seikeigeka Gakkai Zasshi 1995; 69:17-29.

64. Tilg H, Shapiro L, Atkins MB et al. Induction of circulating and erythrocyte-bound IL-8 by IL-2 immunotherapy and suppression of its in vitro production by IL-1 receptor antagonist and soluble tumor necrosis factor receptor (p75) chimera. J Immunol 1993; 151:3299-307.

65. Tilg H, Shapiro L, Vannier E et al. Induction of circulating antagonists to IL-1 and TNF by IL-2 administration and their effects on IL-2 induced cytokine production in vitro. J Immunol 1994; 152:3189-98.

66. Mier JW, Vachino G, Van-der-Meer JW et al. Induction of circulating tumor necrosis factor (TNF alpha) as the mechanism for the febrile response to interleukin-2 (IL-2) in cancer patients. J Clin Immunol 1988; 8:426-36.

67. Numerof RP, Aronson FR, Mier JW. IL-2 stimulates the production of IL-1 alpha and IL-1 beta by human peripheral blood mononuclear cells. J Immunol 1988; 141:4250-7.

68. Martin HA, Murphy PR. Interleukin-2 activates a sub-population of cutaneous C- fibre polymodal nociceptors in the rat hairy skin. Arch Physiol Biochem 1995; 103:136-48.

69. Dinarello CA. The biological properties of interleukin-1. Eur Cytokine Netw 1994; 5:517-31.

70. Rankin EC, Choy EH, Kassimos D et al. The therapeutic effects of an engineered human anti-tumour necrosis factor alpha antibody (CDP571) in rheumatoid arthritis. Br J Rheumatol 1995; 34:334-42.

71. Loetscher H, Gentz R, Zulauf M et al. Recombinant 55-kDa tumor necrosis factor (TNF) receptor. Stoichiometry of binding to TNF alpha and TNF beta and inhibition of TNF activity. J Biol Chem 1991; 266(27):18324-9.

72. Otten U, Gadient RA. Neurotrophins and cytokines—intermediaries between the immune and nervous systems. Int J Dev Neurosci 1995; 13:147-151.

73. Lewin GR, Mendell LM. Nerve growth factor and nociception. Trends Neurosci 1993; 16:353-9.

74. Goedert M, Stoeckel K, Otten U. Biological importance of the retrograde axonal transport of nerve growth factor in sensory neurons. Proc Natl Acad Sci USA 1981; 78:5895-8.

75. Lewin GR, Rueff A, Mendell LM. Peripheral and central mechanisms of ngf-induced hyperalgesia. Eur J Neurosci 1994; 6:1903-12.

76. Andreev NY, Dimitrieva N, Koltzenburg M et al. Peripheral administration of nerve growth factor in the adult rat produces a thermal hyperalgesia that requires the presence of sympathetic post-ganglionic neurones. Pain 1995; 63:109-15.

77. Zurawski SM, Chomarat P, Djossou O et al. The primary binding sub-

unit of the human interleukin-4 receptor is also a component of the interleukin-13 receptor. J Biol Chem 1995; 270:13869-78.

78. Niiro H, Otsuka T, Tanabe T et al. Inhibition by interleukin-10 of inducible cyclooxygenase expression in lipopolysaccharide-stimulated monocytes: its underlying mechanism in comparison with interleukin-4. Blood 1995; 85:3736-45.

79. Seibert K, Zhang Y et al. Pharmacological and biochemical demonstration of the role of cyclooxygenase 2 in inflammation and pain. Proc Natl Acad Sci USA 1994; 91:12013-7.

80. Masferrer JL, Zweifel BS, Manning PT et al. Selective inhibition of inducible cyclooxygenase 2 in vivo is antiinflammatory and nonulcerogenic. Proc Natl Acad Sci USA 1994; 91:3228-32.

81. Tomlinson A, Appleton I, Moore AR, Gilroy DW, Willis D, Mitchell JA, Willoughby DA. Cyclo-oxygenase and nitric oxide synthase isoforms in rat carrageenin-induced pleurisy. Br J Pharmacol 1994; 113:693-8.

82. Chomarat P, Vannier E, Dechanet J et al. Balance of IL-1 receptor antagonist/il-1 beta in rheumatoid synovium and its regulation by IL-4 and IL-10. J Immunol 1995; 154:1432-9.

83. Donnelly RP, Fenton MJ, Kaufman JD et al. IL-1 expression isn human monocytes is transcriptionally and posttranscriptionally regulated by IL-4. J Immunol 1991; 146:3431-6.

84. Wang P, Wu P, Siegel MI, Egan RW et al. Interleukin (Il)-10 inhibits nuclear factor kappa b (nf kappa b) activation in human monocytes— IL-10 and IL-4 suppress cytokine synthesis by different mechanisms. J Biol Chem 1995; 270:9558-63.

85. Dechanet J, Rissoan M, Banchereau J et al. Interleukin 4, but not interleukin 10, regulates the production of inflammation mediators by rheumatoid synoviocytes. Cytokine 1995; 7:176-183.

86. Poole S, Cunha FQ, Selkirk S et al. Cytokine-mediated inflammatory hyperalgesia limited by interleukin-10. Br J Pharmacol 1995; 115:684-8.

87. Dray A, Perkins M. Bradykinin and inflammatory pain. Trends Neurosci 1993; 16:99-104.

88. Bhoola KD, Figueroa CD, Worthy K. Bioregulation of kinins: kallikreins, kininogens, and kininases. Pharmacol Rev 1992; 44(1):1-80.

89. Correa CR, Calixto JB. Evidence for participation of B1 and B2 kinin receptors in formalin-induced nociceptive response in the mouse. Br J Pharmacol 1993; 110:193-8.

90. Perkins MN, Campbell E, Dray A. Antinociceptive activity of the bradykinin B1 and B2 receptor antagonists, desArg9Leu8BK and Hoe 140, in two models of persistent hypealgesia in the rat. Pain 1993; 191-7.

91. Perkins MN, Kelly D. Induction of bradykinin B1 receptors in vivo in a model of ultra-violet irradiation-induced thermal hyperalgesia in the rat. Br J Pharmacol 1993; 110:1441-4.

92. Davis AJ, Perkins MN. Induction of B1 receptors in vivo in a model of persistent inflammatory mechanical hyperalgesia in the rat. Neuropharmacology 1994; 33(1):127-33.

93. Lerner UH, Modeer T. Bradykinin B_1 and B_2 receptor agonists synergistically potentiate interleukin-1-induced prostaglandin biosynthesis in human gingival fibroblasts. Inflammation 1991; 15:427-36.

94. Galizzi JP, Bodinier MC, Chapelain B et al. Up-regulation of (^{3}H)-des-Arg^{10}-kallidin binding to the bradykinin B_1 receptor by interleukin-1beta in isolated smooth muscle cells: correlation with B_1 agonist-induced PGI_2 production. Br J Pharmacol 1994; 113:389-94.

95. Tiffany CW, Burch RM. Bradykinin stimulates tumor necrosis factor and interleukin-1 release from macrophages. FEBS Lett 1989; 247:189-92.

96. Marceau F. Kinin B-1 receptors: a review. Immunopharmacology 1995; 30:1-26.

97. Perretti M, Solito E, Parente L. Evidence that endogenous interleukin-1 is involved in leukocyte migration in acute experimental inflammation in rats and mice. Agents Actions 1992; 35:71-8.

98. Perretti M, Ahluwalia A, Flower RJ, Manzini S. Endogenous tachykinins play a role in IL-1-induced neutrophil accumulation—involvement of NK-1 receptors. Immunology 1993; 80:73-7.

99. Takano M, Yokoyama K, Yayama K et al. Murine fibroblasts synthesize and secrete kininogen in response to cyclic-amp, prostaglandin E(2) and tumor necrosis factor. Bba-Mol Cell Res 1995; 1265:189-95.

100. Rueff A, Mendell LM. NGF-induced thermal hyperalgesia in adult rats involves the activation of bradykinin B1 receptors. Am Soc Neurosci 1994; Abs:287.2.

INDEX

DATE DUE

DEC 0 5 1999	